HOLT
PHYSICS

Assessment Item Listing

HOLT, RINEHART AND WINSTON

A Harcourt Classroom Education Company

Austin · New York · Orlando · Atlanta · San Francisco · Boston · Dallas · Toronto · London

Cover Photo: © Lawrence Manning/CORBIS

Cover Design: Jason Wilson

Copyright © by Holt, Rinehart and Winston

ExamView is a registered trademark of FSCreations, Inc.

Printed in the United States of America

1 2 3 4 5 6 095 04 03 02 01 00

CONTENTS

Introduction

The *Holt Physics* Test Generator and Assessment Item Listing

The *Holt Physics* Test Generator consists of a comprehensive bank of more than 3,500 test items and the *ExamView Pro 3.0* Test Builder software, which enables you to produce your own tests based on these items and those items you create yourself. Both Macintosh® and Windows® versions of the Test Generator are included on the *Holt Physics One-Stop Planner CD-ROM with Test Generator*. Directions on pp. v-vii of this book explain how to install the program on your computer. This Assessment Item Listing is a print-out of all the test items in the Test Generator.

ExamView Pro 3.0 Software

The *ExamView Pro 3.0* Test Builder program enables you to quickly create printed tests. You can enter your own questions and customize the appearance of the tests you create. The *ExamView Pro 3.0* Test Builder program offers many unique features and provides numerous options that allow you to customize the content and appearance of the tests you create.

Test Items

The *Holt Physics* Test Generator contains a file of test items for each chapter of the textbook. The test items are in a variety of formats, including multiple choice, completion, problem, and short answer or essay. Each item is correlated to the chapter objectives in the textbook and by difficulty level.

Item Codes

As you browse through this Assessment Item Listing, you will see that all test items of the same type are gathered together under an identifying head. Each item is coded to assist you with item selection. Following is an explanation of the codes.

MULTIPLE CHOICE

23. A car goes forward along a level road at constant velocity. The force needed to bring the car into equilibrium is
   a. greater than the normal force times the coefficient of static friction
   b. equal to the normal force times the coefficient of static friction
   c. the normal force times the coefficient of static friction
   d. zero

   ANS: D   DIF: I   OBJ: 4-2.3

DIF defines the difficulty of the item.
  I requires recall of information.
  II requires conceptual understanding of known information.
  IIIA requires analysis and interpretation to solve a quantitative problem.
  IIIB often involve solving multiple-step quantitative problems.
  IIIC are challenging problems integrating conceptual understanding and rigorous quantitative problem solving.

OBJ lists the chapter number, section number, and objective. (4-2.3 = Chapter 4, Section 2, Objective 3)

INSTALLATION AND STARTUP

The ExamView Pro 3.0 test generator software is provided on the One-Stop Planner CD-ROM. The ExamView test generator includes the program and all of the questions for the corresponding textbook. Your ExamView software includes three components: Test Builder, Question Bank Editor, and Test Player. The **Test Builder** includes options to create, edit, print, and save tests. The **Question Bank Editor** lets you create or edit question banks. The **Test Player** is a separate program that your students can use to take on-line* (computerized or LAN-based) tests. Please refer to the ExamView User's Guide on the One-Stop Planner CD-ROM for complete instructions.

Before you can use the test generator, you must install the program and the test banks on your hard drive. The system requirements, installation instructions, and startup procedures are provided below.

SYSTEM REQUIREMENTS

To use the ExamView Pro 3.0 test generator, your computer must meet or exceed the following minimum hardware requirements:

Windows®
- Pentium computer
- Windows 95®, Windows 98®, Windows 2000® (or a more recent version)
- color monitor (VGA-compatible)
- CD-ROM and/or high-density floppy disk drive
- hard drive with at least 7 MB space available
- 8 MB available memory (*16 MB memory recommended*)
- *If you wish to use the on-line test player, you must have an Internet connection to access the Internet testing features.**

Macintosh®
- PowerPC processor, 100 MHz computer
- System 7.5 (or a more recent version)
- color monitor (VGA-compatible)
- CD-ROM and/or high-density floppy disk drive
- hard drive with at least 7 MB space available
- 8 MB available memory (*16 MB memory recommended*)
- *If you wish to use the on-line test player, you must have an Internet connection with System 8.6 (or more recent version) to access the Internet testing features.**

You can use the on-line test player to host tests on your personal or school Web site or local area network (LAN) at no additional charge. The ExamView Web site's Internet test-hosting service must be purchased separately. Visit www.examview.com to learn more.

INSTALLATION

Instructions for installing ExamView from the CD-ROM:

Windows®
Step 1
Turn on your computer.
Step 2
Insert the One-Stop Planner disc with the ExamView test generator into the CD-ROM drive.
Step 3
Click the **Start** button on the *Taskbar,* and choose the *Run* option.
Step 4
The ExamView software is provided on the One-Stop Planner CD-ROM under the drive letter that corresponds to the CD-ROM drive on your computer (e.g., **d:\setup.exe**). The setup program will automatically install everything you need to use ExamView.
Step 5
Follow the prompts on the screen to complete the installation process.

Macintosh®
Step 1
Turn on your computer.
Step 2
Insert the One-Stop Planner disc with the ExamView test generator into the CD-ROM drive.
Step 3
Double-click the *ExamView* installation icon to start the program.
Step 4
Follow the prompts on the screen to complete the installation process.

Instructions for installing ExamView from the One-Stop Planner Menu (Macintosh® or Windows®):

Follow steps 1 and 2 from above.
Step 3
Click on **One-Stop.pdf.** (If you do not have Adobe Acrobat® Reader installed on your computer, install it before proceeding by clicking on **Reader Installer.**)
Step 4
Click on the **Test Generator** button.
Step 5
Click on **Install ExamView.**
Step 6
Select the operating system you are using (Macintosh® or Windows®).
Step 7
ExamView will launch the installer. Follow the prompts on the screen to complete the installation process.

GETTING STARTED

After you complete the installation process, follow these instructions to start the ExamView test generator software. See the ExamView User's Guide on the One-Stop Planner for further instructions on the options for creating a test and editing a question bank.

Startup Instructions

Step 1
Turn on the computer.
Step 2
Windows®: Click the **Start** button on the *Taskbar*. Highlight the **Programs** menu, and locate the *ExamView Test Generator* folder. Select the *ExamView Pro* option to start the software.
Macintosh®: Locate and open the *ExamView* folder. Double-click the *ExamView Pro* program icon.
Step 3
The first time you run the software, you will be prompted to enter your name, school/institution name, and city/state. You are now ready to begin using the ExamView software.
Step 4
Each time you start ExamView, the **Startup** menu appears. Choose one of the options shown.
Step 5
Use ExamView to create a test or edit questions in a question bank.

Technical Support

If you have any questions about the Test Generator or need assistance, call the Holt, Rinehart and Winston technical support line at 1-800-323-9239, Monday through Friday, 7:00 A.M. to 6:00 P.M., Central Standard Time. You can contact the Technical Support Center on the Internet at http://www.hrwtechsupport.com or by e-mail at tsc@hrwtechsupport.com.

MULTIPLE CHOICE

1. Which of the following is an area of physics that studies motion and its causes?
   a. thermodynamics
   b. mechanics
   c. quantum mechanics
   d. optics

   ANS: B   DIF: I   OBJ: 1-1.1

2. Which of the following is an area of physics that studies heat and temperature?
   a. thermodynamics
   b. mechanics
   c. quantum mechanics
   d. optics

   ANS: A   DIF: I   OBJ: 1-1.1

3. Listening to your favorite radio station involves which area of physics?
   a. optics
   b. thermodynamics
   c. vibrations and wave phenomena
   d. relativity

   ANS: C   DIF: I   OBJ: 1-1.1

4. A baker makes a loaf of bread. Identify the area of physics that this involves.
   a. optics
   b. thermodynamics
   c. mechanics
   d. relativity

   ANS: B   DIF: I   OBJ: 1-1.1

5. A hiker uses a compass to navigate through the woods. Identify the area of physics that this involves.
   a. thermodynamics
   b. relativity
   c. electromagnetism
   d. quantum mechanics

   ANS: C   DIF: I   OBJ: 1-1.1

6. According to the scientific method, why does a physicist make observations and collect data?
   a. to decide which parts of a problem are important
   b. to ask a question
   c. to make a conclusion
   d. to solve all problems

   ANS: B   DIF: I   OBJ: 1-1.2

7. According to the scientific method, how does a physicist formulate and objectively test hypotheses?
   a. by defending an opinion
   b. by interpreting graphs
   c. by experiments
   d. by stating conclusions

   ANS: C   DIF: I   OBJ: 1-1.2

8. In the steps of the scientific method, what is the next step after formulating and objectively testing hypotheses?
   a. interpreting results
   b. stating conclusions
   c. conducting experiments
   d. making observations and collecting data

   ANS: A   DIF: I   OBJ: 1-1.2

9. According to the scientific method, how should conclusions be stated?
   a. so that no one can refute the conclusion
   b. so that it works with only one set of data
   c. so that it is completely correct, with no mistakes
   d. in a form that can be evaluated by others

   ANS: D   DIF: I   OBJ: 1-1.2

10. Diagrams are NOT designed to
    a. show relationships between concepts.
    b. show setups of experiments.
    c. measure an event or situation.
    d. label parts of a model.

    ANS: C   DIF: I   OBJ: 1-1.3

11. Why do physicists use models?
    a. to explain the complex features of simple phenomena
    b. to describe all aspects of a phenomenon
    c. to explain the basic features of complex phenomena
    d. to describe all of reality

    ANS: C   DIF: I   OBJ: 1-1.3

12. Which statement about models is NOT correct?
    a. Models describe only part of reality.
    b. Models help build hypotheses.
    c. Models help guide experimental design.
    d. Models manipulate a single variable or factor in an experiment.

    ANS: D   DIF: I   OBJ: 1-1.3

13. What two dimensions, in addition to mass, are commonly used by physicists to derive additional measurements?
    a. length and width
    b. area and mass
    c. length and time
    d. velocity and time

    ANS: C   DIF: I   OBJ: 1-2.1

14. The symbol mm represents a
    a. micrometer.
    b. millimeter.
    c. megameter.
    d. manometer.

    ANS: B   DIF: I   OBJ: 1-2.1

15. The symbols for units of length in order from smallest to largest are
    a. m, cm, mm, and km.
    b. mm, m, cm, and km.
    c. km, mm, cm, and m.
    d. mm, cm, m, and km.

    ANS: D   DIF: I   OBJ: 1-2.1

16. The SI base unit used to measure mass is the
    a. meter.
    b. second.
    c. kilogram.
    d. liter.

    ANS: C   DIF: I   OBJ: 1-2.1

17. The SI base unit for time is
    a. 1 day.
    b. 1 hour.
    c. 1 minute.
    d. 1 second.

    ANS: D   DIF: I   OBJ: 1-2.1

18. The most appropriate SI unit for measuring the length of an automobile is the
    a. centimeter.
    b. kilometer.
    c. meter.
    d. millimeter.

    ANS: C   DIF: II   OBJ: 1-2.1

19. If some measurements agree closely with each other but differ widely from the actual value, these measurements are
    a. neither precise nor accurate.
    b. accurate but not precise.
    c. acceptable as a new standard of accuracy.
    d. precise but not accurate.

    ANS: D   DIF: I   OBJ: 1-2.3

20. Poor precision in scientific measurements may arise from
    a. significant figures.
    b. human error.
    c. scientific notation.
    d. both significant figures and scientific notation.

    ANS: B   DIF: I   OBJ: 1-2.3

21. These values were obtained as the mass of a bar of metal: 8.83 g; 8.84 g; 8.82 g. The known mass is 10.68 g. The values are
    a. accurate.
    b. precise.
    c. both accurate and precise.
    d. neither accurate nor precise.

    ANS: B   DIF: II   OBJ: 1-2.3

22. Five darts strike near the center of a target. The dart thrower is
    a. accurate.
    b. precise.
    c. both accurate and precise.
    d. neither accurate nor precise.

    ANS: C   DIF: II   OBJ: 1-2.3

23. In a game of horseshoes, one horseshoe lands on the post. Four horseshoes land nowhere near the post. The one horseshoe on the post was thrown
   a.   accurately.
   b.   precisely.
   c.   both accurately and precisely.
   d.   neither accurately nor precisely.

   ANS: A   DIF: I   OBJ: 1-2.3

24. Calculate the following, and express the answer in scientific notation with the correct number of significant figures: $21.4 + 15 + 17.17 + 4.003$
   a.   57.573
   b.   57.57
   c.   57.6
   d.   58

   ANS: D   DIF: IIIA   OBJ: 1-2.4

25. Calculate the following, and express the answer in scientific notation with the correct number of significant figures: $10.5 \times 8.8 \times 3.14$
   a. 2.9×10^2
   b.   290.136
   c.   290.1
   d.   290

   ANS: A   DIF: IIIA   OBJ: 1-2.4

26. Calculate the following, and express the answer in scientific notation with the correct number of significant figures: $(0.82 + 0.042) \times (4.4 \times 10^3)$
   a. 3.8×10^3
   b. 3.78×10^3
   c. 3.784×10^3
   d.   3784

   ANS: A   DIF: IIIA   OBJ: 1-2.4

Hour	Temperature (°C)
1:00	30.0
2:00	29.0
3:00	28.0
4:00	27.5
5:00	27.0
6:00	25.0

27. A weather balloon records the temperature every hour. From the table above, the temperature
   a.   increases.
   b.   decreases.
   c.   remains constant.
   d.   decreases and then increases.

   ANS: B   DIF: II   OBJ: 1-3.1

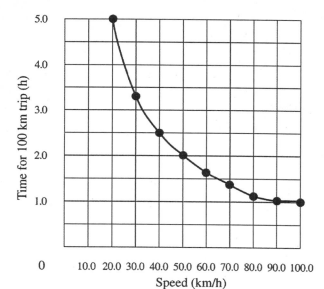

28. The time required to make a trip of 100.0 km is measured at various speeds. From the graph above, what speed will allow the trip to be made in 2 hours?
   a. 20.0 km/h
   b. 40.0 km/h
   c. 50.0 km/h
   d. 90.0 km/h

ANS: C   DIF: II   OBJ: 1-3.1

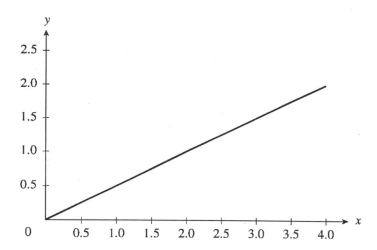

29. Which of the following equations best describes the graph above?
   a. $y = 2x$
   b. $y = x$
   c. $y = x^2$
   d. $y = \frac{1}{2}x$

ANS: D   DIF: II   OBJ: 1-3.1

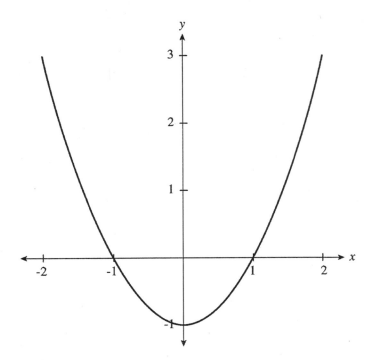

30. Which of the following equations best describes the graph above?
    a. $y = x^2 + 1$
    b. $y = x^2 - 1$
    c. $y = -x^2 + 1$
    d. $y = -x^2 - 1$

    ANS: B   DIF: II   OBJ: 1-3.1

31. The Greek letter *delta*, Δ, indicates a(n)
    a. difference or change.
    b. sum or total.
    c. direct proportion.
    d. inverse proportion.

    ANS: A   DIF: I   OBJ: 1-3.2

32. The Greek letter *sigma*, Σ, indicates a(n)
    a. difference or change.
    b. sum or total.
    c. direct proportion.
    d. inverse proportion

    ANS: B   DIF: I   OBJ: 1-3.2

33. What is the symbol for a time interval?
    a. t
    b. *t*
    c. T
    d. Δt

    ANS: D   DIF: II   OBJ: 1-3.2

34. What is the symbol for mass?
    a. m
    b. *m*
    c. M
    d. Δm

    ANS: B   DIF: II   OBJ: 1-3.2

35. What are the basic SI units?
    a. meters, kilograms, hours
    b. feet, pounds, seconds
    c. meters, kilograms, seconds
    d. feet, kilograms, seconds

    ANS: C   DIF: I   OBJ: 1-3.2

36. Which expression has the same dimensions as an expression yielding a value for acceleration (m/s^2)? (Δv has units of m/s.)
    a. $\Delta v / \Delta t^2$
    b. $\Delta v / \Delta x^2$
    c. $\Delta v^2 / \Delta t$
    d. $\Delta v^2 / \Delta x$

    ANS: D   DIF: IIIA   OBJ: 1-3.3

37. Which expression has the same dimensions as an expression yielding a value for time? (v has units of m/s.)
    a. $\Delta x / v$
    b. $\Delta x / v^2$
    c. $\Delta t / \Delta x$
    d. $1 / v^2 \bullet \Delta t$

    ANS: A   DIF: IIIA   OBJ: 1-3.3

38. Which of the following expressions gives units of kg$\bullet$ m^2/s^2?
    a. $m^2 \bullet \Delta x / \Delta t^2$
    b. $m \bullet \Delta x^2 / \Delta t^2$
    c. $m \bullet \Delta x^2 / \Delta t$
    d. $\Delta t^2 / m \bullet \Delta x^2$

    ANS: B   DIF: IIIA   OBJ: 1-3.3

39. If the change in position Δx is related to velocity v (with units of m/s) in the equation $\Delta x = Av$, the constant A has which dimension?
    a. m/s^2
    b. m
    c. s
    d. m^2

    ANS: C   DIF: II   OBJ: 1-3.3

40. If a is acceleration (m/s^2), Δv is change in velocity (m/s), Δx is change in position (m), and Δt is the time interval (s), which equation is NOT dimensionally correct?
    a. $\Delta t = \Delta x / v$
    b. $a = v^2 / \Delta x$
    c. $\Delta v = a / \Delta t$
    d. $\Delta t^2 = 2 \Delta x / a$

    ANS: D   DIF: IIIA   OBJ: 1-3.3

41. Which of the following equations gives units of s^2? (Δv has units of m/s)
    a. $\Delta v^2 / \Delta x^2$
    b. $m \bullet \Delta v \bullet \Delta t^2 / m$
    c. $\Delta x^2 / \Delta v^2$
    d. $m \bullet \Delta t^2 / m \bullet \Delta v \bullet \Delta t^2$

    ANS: C   DIF: II   OBJ: 1-3.3

42. Estimate the order of magnitude of the length of a football field.
    a. 10^{-1} m
    b. 10^2 m
    c. 10^4 m
    d. 10^6 m

    ANS: B   DIF: II   OBJ: 1-3.4

43. Estimate the order of magnitude of your age, measured in units of months.
    a. 10^{-1} months
    b. 10^1 month
    c. 10^2 months
    d. 10^3 months

    ANS: C   DIF: II   OBJ: 1-3.4

44. The sun is composed mostly of hydrogen. The mass of the sun is 2.0×10^{30} kg, and the mass of a hydrogen atom is 1.67×10^{-27} kg. Estimate the number of atoms in the sun.
    a. 10^3
    b. 10^{57}
    c. 10^{30}
    d. 10^{75}

    ANS: B   DIF: IIIB   OBJ: 1-3.4

SHORT ANSWER

1. List the steps of the scientific method.

   ANS:
   The steps in the scientific method include making observations and collecting data that lead to a question, formulating and objectively testing hypotheses by experiments, interpreting results and revising hypotheses if necessary, and stating conclusions in a form that can be evaluated by others.

   DIF: I   OBJ: 1-1.2

2. How can only seven basic units serve to express almost any measure quantity?

   ANS:
   The basic units can be combined to form units for other quantities.

   DIF: II   OBJ: 1-2.1

3. Convert 92×10^3 km to decimeters using scientific notation.

   ANS:
   9.2×10^8 dm

   DIF: IIIA   OBJ: 1-2.2

4. Distinguish between precision and accuracy.

   ANS:
   Precision is the degree of exactness or refinement of a measurement. Accuracy is the extent to which a reported measurement approaches the true value of the quantity measured.

   DIF: I   OBJ: 1-2.3

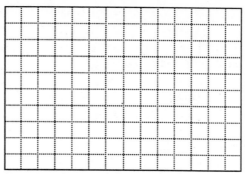

Speedometer reading (km/h)	Time for 100 km trip (h)
20.0	5.00
30.0	3.33
40.0	2.50
50.0	2.00
60.0	1.67
70.0	1.43
80.0	1.25
90.0	1.11
100.0	1.00

5. Using the table above, construct a graph of the time required to make a trip of 100 km measured at various speeds.

ANS:

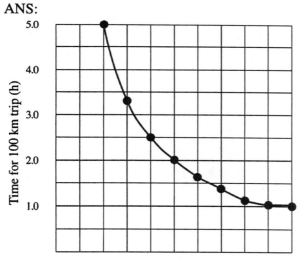

DIF: II   OBJ: 1-3.1

PROBLEM

1. Convert 1 μm to meters using scientific notation.

    ANS:
    1×10^{-6} m

    DIF: IIIA   OBJ: 1-2.2

2. Convert 5.52×10^8 g to kilograms using scientific notation.

    ANS:
    5.52×10^5 kg

    DIF: IIIA   OBJ: 1-2.2

3. Convert 8.66×10^{-9} m to millimeters using scientific notation.

    ANS:
    8.66×10^{-6} mm

    DIF: IIIA   OBJ: 1-2.2

4. Calculate the following, expressing the answer in scientific notation with the correct number of significant figures: $(8.86 + 1.0 \times 10^{-3}) \div 3.610 \times 10^{-3}$

    ANS:
    2.5×10^3

    DIF: IIIA   OBJ: 1-2.4

5. What is the speed of a proton that travels 3.022×10^{12} m in 2.02×10^4 s? Express the answer in scientific notation using the correct number of significant digits.

    ANS:
    1.50×10^8 m/s

    DIF: IIIA   OBJ: 1-2.4

Holt Physics Assessment Item Listing
10

MULTIPLE CHOICE

1. What is the speed of an object at rest?
   a.   0.0 m/s
   b.   1.0 m/s
   c.   9.8 m/s
   d.   9.81 m/s

   ANS: A   OBJ: 2-1.1

2. Which of the following is the expression for average velocity?

   a. $v_{avg} = \dfrac{\Delta x}{\Delta t}$

   b. $v_{avg} = \dfrac{\Delta t}{\Delta x}$

   c. $v_{avg} = \Delta x \bullet \Delta t$

   d. $v_{avg} = \dfrac{v_i + v_f}{2}$

   ANS: D   OBJ: 2-1.1

3. In addition to displacement, which of the following must be used for a more complete description of the average velocity of an object?
   a. m
   b.   kg
   c. Δt
   d. Δx

   ANS: C   OBJ: 2-1.1

4. A dolphin swims 1.85 km/h. How far has the dolphin traveled after 0.60 h?
   a.   1.1 km
   b.   2.5 km
   c.   0.63 km
   d.   3.7 km

   ANS: A   OBJ: 2-1.2

5. A hiker travels south along a straight path for 1.5 h with an average velocity of 0.75 km/h, then travels south for 2.5 h with an average velocity of 0.90 km/h. What is the hiker's displacement for the total trip?
   a.   1.1 km to the south
   b.   2.2 km to the south
   c.   3.4 km to the south
   d.   6.7 km to the south

   ANS: C   OBJ: 2-1.2

6. Acceleration is
   a.   displacement.
   b.   the rate of change of displacement.
   c.   velocity.
   d.   the rate of change of velocity.

   ANS: D   OBJ: 2-2.1

Holt Physics Assessment Item Listing
11

7. Which of the following is the expression for acceleration?

a. $a = \dfrac{\Delta t}{\Delta v}$

c. $a = \Delta t \bullet \Delta v$

b. $a = \dfrac{\Delta v}{\Delta t}$

d. $a = \dfrac{v_i - v_f}{t_i - t_f}$

ANS: B   OBJ: 2-2.1

8. When velocity is positive and acceleration is negative, what happens to the object's motion?
   a. The object slows down.
   b. The object speeds up.
   c. Nothing happens to the object.
   d. The object remains at rest.

ANS: A   OBJ: 2-2.1

9. What does the graph above illustrate about acceleration?
   a. The acceleration is constant.
   b. The acceleration is zero.
   c. The acceleration decreases.
   d. There is not enough information to answer.

ANS: A   OBJ: 2-2.2

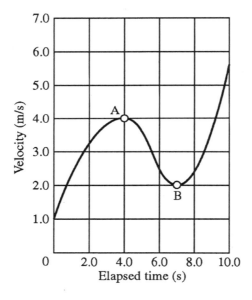

10. What does the graph above illustrate about acceleration?
    a. The acceleration varies.
    b. The acceleration is zero.
    c. The acceleration is constant.
    d. The acceleration increases then becomes constant.

ANS: A   OBJ: 2-2.2

11. A toy car is given an initial velocity of 5.0 m/s and experiences a constant acceleration of 2.0 m/s^2. What is the final velocity after 6.0 s?
    a. 10.0 m/s
    b. 12 m/s
    c. 16 m/s
    d. 17 m/s

    ANS: D   OBJ: 2-2.3

12. A shopping cart given an initial velocity of 2.0 m/s undergoes a constant acceleration of 3.0 m/s^2. What is the magnitude of the cart's displacement after the first 4.0 s of its motion?
    a. 10.0 m
    b. 55 m
    c. 32 m
    d. 80.0 m

    ANS: C   OBJ: 2-2.3

13. A race car accelerates from 0 m/s to 30.0 m/s with a displacement of 45.0 m. What is the vehicle's acceleration?
    a. 2.00 m/s^2
    b. 5.00 m/s^2
    c. 10.0 m/s^2
    d. 15.0 m/s^2

    ANS: C   OBJ: 2-2.3

14. A marble accelerates from rest at a constant rate and travels for a total displacement of 44 m in 20.0 s. What is the average velocity of the marble?
    a. 1.1 m/s
    b. 2.2 m/s
    c. 4.4 m/s
    d. 0.0 m/s

    ANS: A   OBJ: 2-2.3

15. A soccer ball is kicked horizontally. What is its average speed if its displacement is 21.0 m after 4.00 s?
    a. 5.25 m/s
    b. 8.75 m/s
    c. 14.4 m/s
    d. 2.63 m/s

    ANS: D   OBJ: 2-2.3

16. A curious kitten pushes a ball of yarn at rest with its nose, displacing the ball of yarn 17.5 cm in 2.00 s. What is the acceleration of the ball of yarn?
    a. 11.0 cm/s^2
    b. 8.75 cm/s^2
    c. 14.4 cm/s^2
    d. 4.38 cm/s^2

    ANS: D   OBJ: 2-2.3

17. A sports car accelerates at a constant rate from rest to a speed of 27.8 m/s in 8.00 s. What is the displacement of the sports car in this time interval?
    a. 55.0 m
    b. 77.0 m
    c. 111 m
    d. 222 m

    ANS: C   OBJ: 2-2.3

18. Which of the following units are used to measure free fall?
    a. m/s
    b. m/s^2
    c. m•s
    d. m^2/s^2

    ANS: B   OBJ: 2-3.1

19. Which of the following is a value for the acceleration of objects in free fall?
    a. 9.81 m/s^2
    b. −9.81 m/s^2
    c. 9.80 m/s^2
    d. −9.80 m/s^2

    ANS: B   OBJ: 2-3.1

20. Acceleration due to gravity is also called
    a. negative velocity.
    b. displacement.
    c. free-fall acceleration.
    d. instantaneous velocity.

    ANS: C   OBJ: 2-3.1

21. The baseball catcher throws a ball vertically upward and catches it in the same spot as it returns to the mitt. At what point in the ball's path does it experience zero velocity and zero acceleration at the same time?
    a. midway on the way up
    b. at the top of its trajectory
    c. the instant it leaves the catcher's hand
    d. the instant before it arrives in the catcher's mitt

    ANS: B   OBJ: 2-3.1

22. A rock is thrown straight upward with an initial velocity of 24.5 m/s where the downward acceleration due to gravity is 9.81 m/s^2. What is the rock's displacement after 1.00 s?
    a. 9.81 m
    b. 19.6 m
    c. 24.5 m
    d. 29.4 m

    ANS: B   OBJ: 2-3.2

23. A rock is thrown straight upward with an initial velocity of 19.6 m/s where the downward acceleration due to gravity is 9.81 m/s^2. What time interval elapses between the rock's being thrown and its return to the original launch point?
    a. 4.00 s
    b. 5.00 s
    c. 8.00 s
    d. 10.0 s

    ANS: A   OBJ: 2-3.2

24. A baseball is released at rest from the top of the Washington Monument. It hits the ground after falling for 6.00 s. What was the height from which the ball was dropped? (Disregard air resistance. $g = 9.81$ m/s^2.)
    a. 150.0 m
    b. 177 m
    c. 115 m
    d. 210.0 m

    ANS: B   OBJ: 2-3.2

25. A coin released at rest from the top of a tower hits the ground after falling 1.5 s. What is the speed of the coin as it hits the ground? (Disregard air resistance. $g = 9.81$ m/s^2.)
   a.　15 m/s
   b.　21 m/s
   c.　31 m/s
   d.　39 m/s

   ANS: A   OBJ: 2-3.2

26. A rock is thrown from the top of a cliff with an initial speed of 12 m/s. If the rock hits the ground after 2.0 s, what is the height of the cliff? (Disregard air resistance. $g = 9.81$ m/s^2.)
   a.　22 m
   b.　24 m
   c.　44 m
   d.　63 m

   ANS: C   OBJ: 2-3.2

27. A tourist accidentally drops a camera from a 40.0 m high bridge. If $g = 9.81$ m/s^2 and air resistance is disregarded, what is the speed of the camera as it hits the water?
   a.　28.0 m/s
   b.　31.0 m/s
   c.　56.0 m/s
   d.　784 m/s

   ANS: A   OBJ: 2-3.2

28. Human reaction time is usually about 0.20 s. If your lab partner holds a ruler between your finger and thumb and releases it without warning, how far can you expect the ruler to fall before you catch it? (Disregard air resistance. $g = 9.81$ m/s^2.)
   a.　at least 4.0 cm
   b.　at least 9.8 cm
   c.　at least 16.0 cm
   d.　at least 19.6 cm

   ANS: D   OBJ: 2-3.2

29. When there is no air resistance, objects of different masses
   a.　fall with equal accelerations with similar displacements.
   b.　fall with different accelerations with different displacements.
   c.　fall with equal accelerations with different displacements.
   d.　fall with different accelerations with similar displacements.

   ANS: A   OBJ: 2-3.3

30. Objects that are falling toward Earth move
   a.　faster and faster.
   b.　slower and slower.
   c.　at a constant velocity.
   d.　slower then faster.

   ANS: A   OBJ: 2-3.3

31. Which would hit the ground first if dropped from the same height in a vacuum, a feather or a metal bolt?
   a.　the feather
   b.　the metal bolt
   c.　They would hit the ground at the same time.
   d.　They would be suspended in a vacuum.

   ANS: C   OBJ: 2-3.3

32. Which would fall with greater acceleration in a vacuum, a leaf or a stone?
    a. the leaf
    b. the stone
    c. They would accelerate at the same rate.
    d. It is difficult to determine without more information.

    ANS: C   OBJ: 2-3.3

SHORT ANSWER

1. Distinguish between the displacement of a traveler who takes a train from New York to Boston and the displacement of a traveler who flies from Boston to New York. Be sure to compare the magnitudes of the displacements.

    ANS:
    Although the magnitudes of the displacements are equal, the displacements are in opposite directions. Therefore, one displacement is positive and one displacement is negative.

    OBJ: 2-1.1

2. What is free fall?

    ANS:
    Free fall is the motion of an object falling with a constant acceleration in the absence of air resistance.

    OBJ: 2-3.1

3. Why is the direction of free-fall acceleration negative?

    ANS:
    The direction of free-fall acceleration is negative because the object accelerates toward Earth (the usual choice of coordinates uses positive as the direction away from Earth).

    OBJ: 2-3.1

4. What is the acceleration of an object thrown upward? What is its acceleration as it free falls?

    ANS:
    The acceleration is a constant -9.81 m/s^2 throughout its time in the air.

    OBJ: 2-3.1

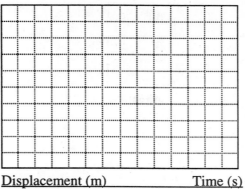

Displacement (m)	Time (s)
1.0	2.0
2.0	4.0
3.0	6.0
4.0	8.0
5.0	10.0

5. Construct a graph of position versus time using the data in the table above. What value is represented by the slope of a graph? Find the slope between $\Delta t = 1$ s and $\Delta t = 2$ s. Be sure to use appropriate SI units.

ANS:
velocity, 0.5 m/s

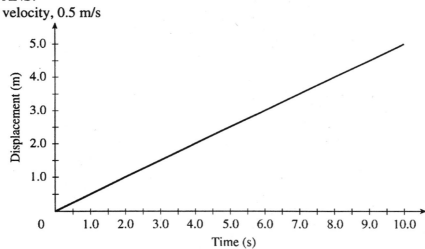

OBJ: 2-1.3

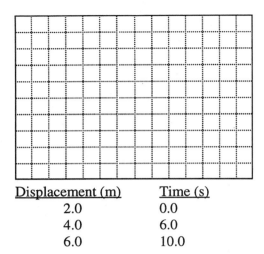

Displacement (m)	Time (s)
2.0	0.0
4.0	6.0
6.0	10.0

6. Construct a graph of position versus time using the data in the table above. What value is represented by the slope of a graph? Find the slope between $\Delta t = 1$ s and $\Delta t = 2$ s. Be sure to use appropriate SI units.

ANS:
velocity, 0.25 m/s

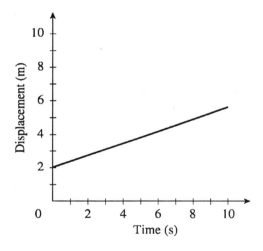

OBJ: 2-1.3

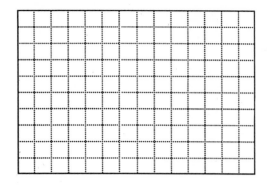

7. A motorized scooter starts from rest and accelerates for 4 s at 2 m/s². It continues at a constant speed for 6 s. Graph velocity versus time to describe this motion.

ANS:

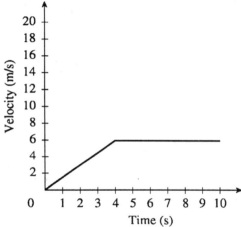

OBJ: 2-2.2

PROBLEM

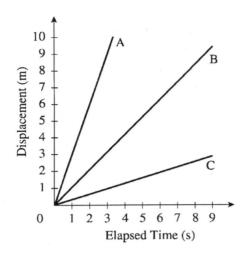

1. The graph above shows displacement versus time. What is the average velocity for line A?

   ANS:
   3.0 m/s

   OBJ: 2-1.3

2. The graph above shows displacement versus time. What is the average velocity for line B?

   ANS:
   1.0 m/s

   OBJ: 2-1.3

3. The graph above shows displacement versus time. What is the average velocity for line C?

   ANS:
   0.3 m/s

   OBJ: 2-1.3

4. A pair of spectacles are dropped from the top of a 32.0 m high stadium. A pen is dropped 2.0 s later. How high above the ground is the pen when the spectacles hit the ground? (Disregard air resistance, $g = 9.81$ m/s^2.)

   ANS:
   3.0×10^1 m

   OBJ: 2-3.3

MULTIPLE CHOICE

1. Which of the following is a physical quantity that has a magnitude but no direction?
   a. vector
   b. scalar
   c. resultant
   d. frame of reference

   ANS: B   DIF: I   OBJ: 3-1.1

2. Which of the following is a physical quantity that has both magnitude and direction?
   a. vector
   b. scalar
   c. resultant
   d. frame of reference

   ANS: A   DIF: I   OBJ: 3-1.1

3. Identify the following quantities as scalar or vector: the mass of an object, the number of leaves on a tree, wind velocity.
   a. vector, scalar, scalar
   b. scalar, scalar, vector
   c. scalar, vector, scalar
   d. vector, scalar, vector

   ANS: B   DIF: II   OBJ: 3-1.1

4. Identify the following quantities as scalar or vector: the speed of a snail, the time it takes to run a mile, the free-fall acceleration.
   a. vector, scalar, scalar
   b. scalar, scalar, vector
   c. vector, scalar, vector
   d. scalar, vector, vector

   ANS: B   DIF: II   OBJ: 3-1.1

5. Which of the following is an example of a vector quantity?
   a. velocity
   b. temperature
   c. volume
   d. mass

   ANS: A   DIF: I   OBJ: 3-1.1

6. For the winter, a duck flies 10.0 m/s due south against a gust of wind with a velocity of 2.5 m/s. What is the resultant velocity of the duck?
   a. 12.5 m/s south
   b. −12.5 m/s south
   c. 7.5 m/s south
   d. −7.5 m/s south

   ANS: C   DIF: II   OBJ: 3-1.2

7. A lightning bug flies at a velocity of 0.25 m/s due east toward another lightning bug seen off in the distance. A light easterly breeze blows on the bug at a velocity of 0.25 m/s. What is the resultant velocity of the lightning bug?
   a. 0.50 m/s
   b. 0.00 m/s
   c. 0.75 m/s
   d. 0.25 m/s

   ANS: B   DIF: II   OBJ: 3-1.2

8. A jogger runs 10.0 blocks due east, 5.0 blocks due south, and another 2.0 blocks due east. Assume all blocks are of equal size. Use the graphical method to find the magnitude of the jogger's net displacement.
   a. 14.0 blocks
   b. 8.0 blocks
   c. 11.0 blocks
   d. 13.0 blocks

   ANS: D   DIF: IIIB   OBJ: 3-1.2

9. A cave explorer travels 3.0 m eastward, then 2.5 m northward, and finally 15 m westward. Use the graphical method to find the magnitude of the net displacement.
   a. 12 m
   b. 5.7 m
   c. 18 m
   d. 15 m

   ANS: A   DIF: IIIB   OBJ: 3-1.2

10. A student adds two vectors with magnitudes of 200 and 40. Taking into account significant figures, which is the only possible choice for the magnitude of the resultant?
    a. 160
    b. 200
    c. 300
    d. 240

    ANS: B   DIF: IIIA   OBJ: 3-1.2

11. A car travels down a road at a certain velocity, v_{car}. The driver slows down so that the car is traveling only half as fast as before. Which of the following is the correct expression for the resulting velocity?
    a. $2v_{car}$
    b. $\frac{1}{2}v_{car}$
    c. $-\frac{1}{2}v_{car}$
    d. $-2v_{car}$

    ANS: B   DIF: II   OBJ: 3-1.3

12. A football player runs in one direction to catch a pass, then turns and runs twice as fast in the opposite direction toward the goal line. Which of the following is the correct expression for the original velocity and the resulting velocity?
    a. $-v_{player}, -2v_{player}$
    b. $v_{player}, 2v_{player}$
    c. $v_{player}, -2v_{player}$
    d. $2v_{player}, -v_{player}$

    ANS: C   DIF: II   OBJ: 3-1.3

13. Multiplying or dividing vectors by scalars results in
    a. vectors.
    b. scalars.
    c. vectors if multiplied or scalars if divided.
    d. scalars if multiplied or vectors if divided.

    ANS: A   DIF: I   OBJ: 3-1.3

14. An airplane flying at 120 km/h due west moves into a region where the wind is blowing at 40 km/h due east. If the plane's original vector velocity is v_{plane}, which of the following is the correct expression for the plane's resulting velocity?

    a. $\frac{2}{3}v_{plane}$ c. $\frac{3}{4}v_{plane}$

    b. $-\frac{1}{3}v_{plane}$ d. $-\frac{2}{3}v_{plane}$

ANS: A   DIF: IIIA   OBJ: 3-1.3

15. A student walks from the door of the house to the end of the driveway and realizes that he missed the bus. The student runs back to the house, traveling three times as fast. Which of the following is the correct expression for the resulting velocity?

    a. $3v_{student}$ c. $\frac{1}{3}v_{student}$

    b. $\frac{1}{3}v_{student}$ d. $-3v_{student}$

ANS: D   DIF: I   OBJ: 3-1.3

16. Which of the following is the best coordinate system to analyze a painter climbing a ladder at an angle of 60° to the ground?
    a. x-axis: horizontal along the ground; y-axis: along the ladder
    b. x-axis: up and down; y-axis: horizontal along the ground
    c. x-axis: horizontal along the ground; y-axis: up and down
    d. x-axis: along the ladder; y-axis: up and down

ANS: C   DIF: I   OBJ: 3-2.1

17. Which of the following is the best coordinate system to analyze a car traveling northeast from one city to another?
    a.   positive x-axis pointing east; positive y-axis pointing south
    b.   positive x-axis pointing west; positive y-axis pointing east
    c.   positive x-axis pointing north; positive y-axis pointing south
    d.   positive x-axis pointing east; positive y-axis pointing north

ANS: D   DIF: I   OBJ: 3-2.1

18. Which of the following is the best coordinate system to analyze an object thrown into the air?
    a. x-axis: perpendicular to the ground; y-axis: up and down
    b. x-axis: up and down; y-axis: parallel to the ground
    c. x-axis: parallel to the ground; y-axis: perpendicular to the ground
    d. x-axis: up and down; y-axis: perpendicular to the ground

ANS: C   DIF: I   OBJ: 3-2.1

19. Which of the following is the best coordinate system to analyze the movement of a submarine diving at an angle of 45° to the surface of the water?
   a. *x*-axis: horizontal at the water level; *y*-axis: up and down
   b. *x*-axis: horizontal at the water level; *y*-axis: left and right
   c. *x*-axis: left and right; *y*-axis: horizontal at the ocean bottom
   d. *x*-axis: up and down; *y*-axis: horizontal at the ocean bottom

   ANS: A   DIF: I   OBJ: 3-2.1

20. An ant on a picnic table travels 3.0×10^1 cm eastward, then 25 cm northward, and finally 15 cm westward. What is the ant's directional displacement relative to its original position?
   a. 29 cm at 59° north of east   c. 57 cm at 29° north of west
   b. 52 cm at 29° north of east   d. 29 cm at 77° north of east

   ANS: A   DIF: IIIB   OBJ: 3-2.2

21. A duck waddles 2.5 m east and 6.0 m north. What are the magnitude and direction of the duck's displacement with respect to its original position?
   a. 3.5 m at 19° north of east   c. 6.5 m at 67° north of east
   b. 6.3 m at 67° north of east   d. 6.5 m at 72° north of east

   ANS: C   DIF: IIIB   OBJ: 3-2.2

22. A quarterback takes the ball from the line of scrimmage and runs backward for 1.0×10^1 m then sideways parallel to the line of scrimmage for 15 m. The ball is thrown forward 5.0×10^1 m perpendicular to the line of scrimmage. The receiver is tackled immediately. How far is the football displaced from its original position?
   a. 43 m   c. 62 m
   b. 55 m   d. 75 m

   ANS: A   DIF: IIIB   OBJ: 3-2.2

23. A plane flies from city A to city B. City B is 1540 km west and 1160 km south of city A. What is the total displacement and direction of the plane?
   a. 1930 km, 43.0° south of west   c. 1850 km, 37.0° south of west
   b. 1850 km, 43.0° south of west   d. 1930 km, 37.0° south of west

   ANS: D   DIF: IIIB   OBJ: 3-2.2

24. While following directions on a treasure map, a person walks 45.0 m south, then turns and walks 7.50 m east. Which single straight-line displacement could the treasure hunter have walked to reach the same spot?
   a. 45.6 m at 9.5° south of east   c. 45.6 m at 9.5° east of south
   b. 52.5 m at 21° east of south   d. 45.6 m at 21° south of east

   ANS: C   DIF: IIIC   OBJ: 3-2.2

25. In a coordinate system, the *x*-component of a given vector is equal to that vector's magnitude multiplied by which trigonometric function, with respect to the angle between the vector and the *x*-axis?
    a. the cosine of θ
    b. the sine of θ
    c. the tangent of θ
    d. the cotangent of θ

    ANS: A   DIF: II   OBJ: 3-2.3

26. In a coordinate system, if the *x* component of a vector and the angle between the vector and *x*-axis are known, then the magnitude of the vector is calculated by which operation, taken with respect to the *x* component?
    a. dividing by the sine of θ
    b. dividing by the cosine of θ
    c. multiplying by the sine of θ
    d. multiplying by the cosine of θ ·

    ANS: B   DIF: II   OBJ: 3-2.3

27. A string attached to an airborne kite was maintained at an angle of 40.0° with the ground. If 120 m of string was reeled in to return the kite back to the ground, what was the horizontal displacement of the kite? (Assume the kite string did not sag.)
    a. 110 m
    b. 84 m
    c. 77 m
    d. 92 m

    ANS: D   DIF: IIIB   OBJ: 3-23

28. An athlete runs 110 m across a level field at an angle of 30.0° north of east. What are the east and north components, respectively, of this displacement?
    a. 64 m; 190 m
    b. 190 m; 64 m
    c. 95 m; 55 m
    d. 55 m; 95 m

    ANS: C   DIF: IIIB   OBJ: 3-2.3

29. A skateboarder rolls 25.0 m down a hill that descends at an angle of 20.0° with the horizontal. Find the horizontal and vertical components of the skateboarder's displacement.
    a. 8.55 m; 23.5 m
    b. 23.5 m; 8.55 m
    c. 23.5 m; 73.1 m
    d. 73.1 m; 26.6 m

    ANS: B   DIF: IIIB   OBJ: 3-2.3

30. Find the resultant of these two vectors: 2.00×10^2 units due east and 4.00×10^2 units 30.0° north of west.
    a. 300 units 29.8° north of west
    b. 581 units 20.1° north of east
    c. 546 units 59.3° north of west
    d. 248 units 53.9° north of west

    ANS: D   DIF: IIIB   OBJ: 3-2.4

31. Vector **A** is 3.2 units in length and points along the positive *y*-axis. Vector **B** is 4.6 units in length and points along a direction 195° counterclockwise from the positive *x*-axis. What is the magnitude of the resultant when vectors **A** and **B** are added?
    a. 1.2 units
    b. 6.2 units
    c. 4.8 units
    d. 5.6 units

    ANS: D   DIF: IIIB   OBJ: 3-2.4

32. What is the resultant displacement of a dog looking for its bone in the yard, if the dog first heads 55° north of west for 10.0 m, and then turns and heads west for 5.00 m?
    a.   11.2 m at 63° west of north
    b.   13.5 m at 37° north of west
    c.   13.5 m at 37° north of east
    d.   62.1 m at 74° north of west

    ANS: B   DIF: II   OBJ: 3-2.4

33. A hiker walks 4.5 km at an angle of 45° north of west. Then the hiker walks 4.5 km south. What is the magnitude and direction of the hiker's total displacement?
    a.   3.5 km, 22° south of west
    b.   3.5 km, 22° north of west
    c.   6.4 km, 45° north of west
    d.   6.4 km, 22° south of west

    ANS: A   DIF: II   OBJ: 3-2.4

34. Which of the following is the motion of objects moving in two dimensions under the influence of gravity?
    a.   horizontal velocity
    b.   directrix
    c.   parabola
    d.   projectile motion

    ANS: D   DIF: I   OBJ: 3-3.1

35. Which of the following is an example of projectile motion?
    a.   a jet lifting off a runway
    b.   a bullet being fired from a gun
    c.   dropping an aluminum can into the recycling bin
    d.   a space shuttle orbiting Earth

    ANS: B   DIF: I   OBJ: 3-3.1

36. Which of the following is NOT an example of projectile motion?
    a.   a volleyball served over a net
    b.   a baseball hit by a bat
    c.   a hot-air balloon drifting toward Earth
    d.   a long jumper in action

    ANS: C   DIF: I   OBJ: 3-3.1

37. What is the path of a projectile?
    a.   a wavy line
    b.   a parabola
    c.   a hyperbola
    d.   Projectiles do not follow a predictable path.

    ANS: B   DIF: I   OBJ: 3-3.2

38. Which of the following exhibits parabolic motion?
    a.   a person diving into a pool from a diving board
    b.   a space shuttle orbiting Earth
    c.   a leaf falling from a tree
    d.   a train moving along a flat track

    ANS: A   DIF: I   OBJ: 3-3.2

39. Which of the following does NOT exhibit parabolic motion?
    a. a frog jumping from land into water
    b. a basketball thrown to a hoop
    c. a flat piece of paper released from a window
    d. a baseball thrown to home plate

    ANS: C   DIF: I   OBJ: 3-3.2

40. A stone is thrown at an angle of 30.0° above the horizontal from the top edge of a cliff with an initial speed of 12 m/s. A stopwatch measures the stone's trajectory time from the top of the cliff to the bottom at 5.6 s. What is the height of the cliff? (Disregard air resistance. $g = 9.81$ m/s².)
    a. 58 m   c. 120 m
    b. 150 m   d. 180 m

    ANS: C   DIF: II   OBJ: 3-3.3

41. A track star in the long jump goes into the jump at 12 m/s and launches herself at 20.0° above the horizontal. How long is she in the air before returning to Earth? ($g = 9.81$ m/s²)
    a. 0.42 s   c. 1.5 s
    b. 0.83 s   d. 1.2 s

    ANS: B   DIF: IIIB   OBJ: 3-3.3

42. A model rocket flies horizontally off the edge of the cliff at a velocity of 50.0 m/s. If the canyon below is 100.0 m deep, how far from the edge of the cliff does the model rocket land?
    a. 112 m   c. 337 m
    b. 225 m   d. 400 m

    ANS: B   DIF: IIIB   OBJ: 3-3.3

43. A firefighter 50.0 m away from a burning building directs a stream of water from a fire hose at an angle of 30.0° above the horizontal. If the velocity of the stream is 40.0 m/s, at what height will the stream of water strike the building?
    a. 9.60 m   c. 18.7 m
    b. 13.4 m   d. 22.4 m

    ANS: C   DIF: IIIB   OBJ: 3-3.3

44. Which of the following is a coordinate system for specifying the precise location of objects in space?
    a. x-axis   c. frame of reference
    b. y-axis   d. diagram

    ANS: C   DIF: II   OBJ: 3-4.1

45. A passenger on a bus moving east sees a man standing on a curb. From the passenger's perspective, the man appears to
    a. stand still.
    b. move west at a speed that is less than the bus's speed.
    c. move west at a speed that is equal to the bus's speed.
    d. move east at a speed that is equal to the bus's speed.

    ANS: C   DIF: I   OBJ: 3-4.1

46. A piece of chalk is dropped by a teacher walking at a speed of 1.5 m/s. From the teacher's perspective, the chalk appears to fall
   a. straight down.
   b. straight down and backward.
   c. straight down and forward.
   d. straight backward.

   ANS: A   DIF: I   OBJ: 3-4.1

47. A jet moving at 500.0 km/h due east moves into a region where the wind is blowing at 120.0 km/h in a direction 30.0° north of east. What is the new velocity and direction of the aircraft relative to the ground?
   a. 607 km/h, 5.67° north of east
   b. 620.0 km/h, 7.10° north of east
   c. 550.0 km/h, 6.22° north of east
   d. 588 km/h, 4.87° north of east

   ANS: A   DIF: IIIB   OBJ: 3-4.2

48. A boat moves at 10.0 m/s relative to the water. If the boat is in a river where the current is 2.00 m/s, how long does it take the boat to make a complete round trip of 1000.0 m upstream followed by 1000.0 m downstream?
   a. 199 s
   b. 203 s
   c. 208 s
   d. 251 s

   ANS: C   DIF: IIIC   OBJ: 3-4.2

49. A superhero flying at treetop level sees the Eiffel Tower elevator begin to free fall. If the superhero is 1.00 km away from the tower and the elevator falls from a height of 240.0 m, how long does the superhero have to save the people in the elevator? What should the superhero's average velocity be?
   a. 7 s; 333 m/s
   b. 5 s; 200 m/s
   c. 7 s; 143 m/s
   d. 9 s; 111 m/s

   ANS: C   DIF: IIIB   OBJ: 3-4.2

50. A small airplane flies at a velocity of 145 km/h toward the south as observed by a person on the ground. The airplane pilot measures an air velocity of 170.0 km/h south. What is the velocity of the wind that affects the plane?
   a. 25 km/h south
   b. 25 km/h north
   c. 315 km/h south
   d. 315 km/h north

   ANS: B   DIF: II   OBJ: 3-4.2

SHORT ANSWER

1. Briefly explain why a basketball being thrown toward the hoop is considered projectile motion.

   ANS:
   Objects sent into the air and subject to gravity exhibit projectile motion.

   DIF: I   OBJ: 3-3.1

2. Briefly explain why the true path of a projectile traveling through Earth's atmosphere is not a parabola.

ANS:
With air resistance, a projectile slows down as it collides with air particles. Therefore, the true path of a projectile would not be a parabola.

DIF: II   OBJ: 3-3.2

PROBLEM

1. A dog walks 24 steps north and then walks 55 steps west to bury a bone. If the dog walks back to the starting point in a straight line, how many steps will the dog take and in which direction will the dog walk? Use the graphical method to find the magnitude of the net displacement.

ANS:
60 steps 24° south of east

DIF: IIIA   OBJ: 3-2.2

MULTIPLE CHOICE

1. Which of the following is the cause of an acceleration or a change in an object's motion?
   a. speed
   b. inertia
   c. force
   d. velocity

   ANS: C   DIF: I   OBJ: 4-1.1

2. Which of the following statements does NOT describe force?
   a. Force causes objects at rest to remain stationary.
   b. Force causes objects to start moving.
   c. Force causes objects to stop moving.
   d. Force causes objects to change direction.

   ANS: A   DIF: I   OBJ: 4-1.1

3. What causes a moving object to change direction?
   a. acceleration
   b. velocity
   c. inertia
   d. force

   ANS: D   DIF: I   OBJ: 4-1.1

4. Which of the following forces arises from direct physical contact between two objects?
   a. gravitational force
   b. fundamental force
   c. contact force
   d. field force

   ANS: C   DIF: I   OBJ: 4-1.2

5. Which of the following forces exists between objects even in the absence of direct physical contact?
   a. frictional force
   b. fundamental force
   c. contact force
   d. field force

   ANS: D   DIF: I   OBJ: 4-1.2

6. Which of the following forces is an example of a contact force?
   a. gravitational force
   b. magnetic force
   c. electric force
   d. frictional force

   ANS: D   DIF: I   OBJ: 4-1.2

7. Which of the following forces is an example of a field force?
   a. gravitational force
   b. frictional force
   c. normal force
   d. tension

   ANS: A   DIF: I   OBJ: 4-1.2

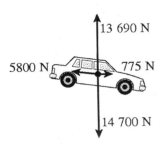

8. In the free-body diagram shown above, which of the following is the gravitational force acting on the car?
   a. 5800 N
   b. 775 N
   c. 14 700 N
   d. 13 690 N

   ANS: C   DIF: I   OBJ: 4-1.3

9. In the free-body diagram show above, the 5800 N force represents
   a. the gravitational force acting on the car.
   b. the backward force the road exerts on the car.
   c. the upward force the road exerts on the car.
   d. the force exerted by a towing cable on the car.

   ANS: D   DIF: I   OBJ: 4-1.3

10. A free-body diagram of a ball in free fall in the presence of air resistance would show
    a. a downward arrow to represent the force of air resistance.
    b. only a downward arrow to represent the force of gravity.
    c. a downward arrow to represent the force of gravity and an upward arrow to represent the force of air resistance.
    d. an upward arrow to represent the force of gravity and a downward arrow to represent the force of air resistance.

    ANS: C   DIF: I   OBJ: 4-1.3

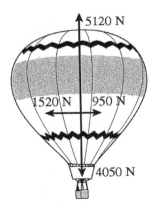

11. In the free-body diagram shown above, which of the following is the gravitational force acting on the balloon?
    a. 1520 N
    b. 950 N
    c. 4050 N
    d. 5120 N

    ANS: C   DIF: I   OBJ: 4-1.3

Holt Physics Assessment Item Listing
31

12. Which of the following is the tendency of an object to maintain its state of motion?
    a. acceleration
    b. inertia
    c. force
    d. velocity

    ANS: B   DIF: I   OBJ: 4-2.1

13. A late traveler rushes to catch a plane, pulling a suitcase with a force directed 30.0° above the horizontal. If the horizontal component of the force on the suitcase is 60.6 N, what is the force exerted on the handle?
    a. 53.0 N
    b. 70.0 N
    c. 65.2 N
    d. 95.6 N

    ANS: B   DIF: IIIB   OBJ: 4-2.2

14. A car goes forward along a level road at constant velocity. The additional force needed to bring the car into equilibrium is
    a. greater than the normal force times the coefficient of static friction.
    b. equal to the normal force times the coefficient of static friction.
    c. the normal force times the coefficient of kinetic friction.
    d. zero.

    ANS: D   DIF: I   OBJ: 4-2.3

15. A sled is pulled at a constant velocity across a horizontal snow surface. If a force of 8.0×10^1 N is being applied to the sled rope at an angle of 53° to the ground, what is the force of friction between the sled and the snow?
    a. 83 N
    b. 64 N
    c. 48 N
    d. 42 N

    ANS: C   DIF: IIIB   OBJ: 4-2.3

16. A trapeze artist weighs 8.00×10^2 N. The artist is momentarily held to one side of a swing by a partner so that both of the swing ropes are at an angle of 30.0° with the vertical. In such a condition of static equilibrium, what is the horizontal force being applied by the partner?
    a. 924 N
    b. 433 N
    c. 196 N
    d. 462 N

    ANS: D   DIF: IIIB   OBJ: 4-2.3

17. If a nonzero net force is acting on an object, then the object is definitely
    a. at rest.
    b. moving with a constant velocity.
    c. being accelerated.
    d. losing mass.

    ANS: C   DIF: I   OBJ: 4-3.1

18. A wagon with a weight of 300.0 N is accelerated across a level surface at 0.5 m/s². What net force acts on the wagon? ($g = 9.81$ m/s²)
    a. 9.0 N
    b. 15 N
    c. 150 N
    d. 610 N

    ANS: B   DIF: IIIB   OBJ: 4-3.1

19. Which statement about the acceleration of an object is correct?
    a. The acceleration of an object is directly proportional to the net external force acting on the object and inversely proportional to the mass of the object.
    b. The acceleration of an object is directly proportional to the net external force acting on the object and directly proportional to the mass of the object.
    c. The acceleration of an object is inversely proportional to the net external force acting on the object and inversely proportional to the mass of the object.
    d. The acceleration of an object is inversely proportional to the net external force acting on the object and directly proportional to the mass of the object.

    ANS: A

20. A small force acting on a human-sized object causes
    a. a small acceleration.   c. a large acceleration.
    b. no acceleration.   d. equilibrium.

    ANS: A

21. According to Newton's second law, when the same force is applied to two objects of different masses,
    a. the object with greater mass will experience a great acceleration and the object with less mass will experience an even greater acceleration.
    b. the object with greater mass will experience a smaller acceleration and the object with less mass will experience a greater acceleration.
    c. the object with greater mass will experience a greater acceleration and the object with less mass will experience a smaller acceleration.
    d. the object with greater mass will experience a small acceleration and the object with less mass will experience an even smaller acceleration.

    ANS: B

22. Two perpendicular forces, one of 45.0 N directed upward and the second of 60.0 N directed to the right, act simultaneously on an object with a mass of 35.0 kg. What is the magnitude of the resultant acceleration of the object?
    a. 8.68 m/s^2
    b. 3.00 m/s^2
    c. 5.25 m/s^2
    d. 1.41 m/s^2

    ANS: A

23. A sailboat with a mass of 2.0×10^3 kg experiences a tidal force of 3.0×10^3 N directed to the east and a wind force against its sails with a magnitude of 6.0×10^3 N directed toward the northwest (45° N of W). What is the magnitude of the resultant acceleration of the boat?
    a. 2.2 m/s^2 c. 1.5 m/s^2
    b. 2.1 m/s^2 d. 4.4 m/s^2

    ANS: A

24. An airplane with a mass of 1.2×10^4 kg tows a glider with a mass of 0.60×10^4 kg. If the airplane propellers provide a net forward thrust of 3.6×10^4 N, what is the acceleration of the glider?
   a. 2.0 m/s^2
   b. 3.0 m/s^2
   c. 6.0 m/s^2
   d. 9.8 m/s^2

   ANS: A   DIF: IIIB   OBJ: 4-3.2

25. An elevator weighing 2.00×10^5 N is supported by a steel cable. What is the tension in the cable when the elevator is accelerated upward at a rate of 3.00 m/s^2? ($g = 9.81$ m/s^2)
   a. 1.39×10^5 N
   b. 2.31×10^5 N
   c. 2.42×10^5 N
   d. 2.61×10^5 N

   ANS: D   DIF: IIIB   OBJ: 4-3.2

26. A hammer drives a nail into a piece of wood. Identify an action-reaction pair, and compare the forces exerted by each object.
   a. The nail exerts a force on the hammer; the hammer exerts a force on the wood.
   b. The hammer exerts a force on the nail; the wood exerts a force on the nail.
   c. The hammer exerts a force on the nail; the nail exerts a force on the hammer.
   d. The hammer exerts a force on the nail; the hammer exerts a force on the wood.

   ANS: C   DIF: I   OBJ: 4-3.3

27. A hockey stick hits a puck on the ice. Identify an action-reaction pair, and compare the forces exerted by each object.
   a. The stick exerts a force on the puck; the puck exerts a force on the stick.
   b. The stick exerts a force on the puck; the puck exerts a force on the ice.
   c. The puck exerts a force on the stick; the stick exerts a force on the ice.
   d. The stick exerts a force on the ice; the ice exerts a force on the puck.

   ANS: A   DIF: I   OBJ: 4-3.3

28. A leaf falls from a tree and lands on the sidewalk. Identify an action-reaction pair, and compare the forces exerted by each object.
   a. The tree exerts a force on the leaf; the sidewalk exerts a force on the leaf.
   b. The leaf exerts a force on the sidewalk; the sidewalk exerts a force on the leaf.
   c. The leaf exerts a force on the tree; the sidewalk exerts a force on the leaf.
   d. The leaf exerts a force on the sidewalk; the tree exerts a force on the leaf.

   ANS: B   DIF: I   OBJ: 4-3.3

29. A ball is dropped from a person's hand and falls to Earth. Identify an action-reaction pair, and compare the forces exerted by each object.
   a. The hand exerts a force on the ball; Earth exerts a force on the hand.
   b. Earth exerts a force on the ball; the hand exerts a force on Earth.
   c. Earth exerts a force on the hand; the hand exerts a force on the ball.
   d. Earth exerts a gravitational force on the ball; the ball exerts a gravitational force on Earth.

   ANS: D   DIF: I   OBJ: 4-3.3

30. The statement by Newton that for every action there is an equal but opposite reaction is which of his laws of motion?
   a. first
   b. second
   c. third
   d. fourth

   ANS: C   DIF: I   OBJ: 4-3.3

31. Which are simultaneous equal but opposite forces resulting from the interaction of two objects?
   a. net external forces
   b. field forces
   c. gravitational forces
   d. action-reaction pairs

   ANS: D   DIF: I   OBJ: 4-3.3

32. As a basketball player starts to jump for a rebound, the player begins to move upward faster and faster until his shoes leave the floor. During the time that the player is in contact with the floor, the force of the floor on the shoes is
   a. greater than the player's weight.
   b. equal in magnitude and opposite in direction to the player's weight.
   c. less than the player's weight.
   d. zero.

   ANS: A   DIF: IIIB   OBJ: 4-3.4

33. The magnitude of the force of gravity acting on an object is
   a. frictional force.
   b. weight.
   c. inertia.
   d. mass.

   ANS: B   DIF: I   OBJ: 4-4.1

34. A measure of the quantity of matter is
   a. density.
   b. weight.
   c. force.
   d. mass.

   ANS: D   DIF: I   OBJ: 4-4.1

35. A change in the force of gravity acting on an object will affect the object's
   a. mass.
   b. frictional force.
   c. weight.
   d. inertia.

   ANS: C   DIF: I   OBJ: 4-4.1

36. A weight of 5.00×10^3 N is suspended in equilibrium by two cables. Cable 1 applies a horizontal force to the right of the object and has a tension, F_{T1}. Cable 2 applies a force upward and to the left at an angle of 37.0° to the negative x-axis and has a tension, F_{T2}. What is F_{T2}?
   a. 4440 N
   b. 6640 N
   c. 8310 N
   d. 3340 N

   ANS: C   DIF: IIIB   OBJ: 4-4.2

37. A sled weighing 1.0×10^2 N is held in place on a frictionless 20.0° slope by a rope attached to a stake at the top. The rope is parallel to the slope. What is the normal force of the slope acting on the sled?

a. 94 N
b. 47 N
c. 37 N
d. 34 N

ANS: A   DIF: IIIB   OBJ: 4-4.2

38. A mule uses a rope to pull a box that weighs 3.0×10^2 N across a level surface with constant velocity. The rope makes an angle of 30.0° above the horizontal, and the tension in the rope is 1.0×10^2 N. What is the normal force of the floor on the box?

a. 300.0 N
b. 86 N
c. 50.0 N
d. 250 N

ANS: D   DIF: IIIB   OBJ: 4-4.2

39. A book with a mass of 2.0 kg is held in equilibrium on a board with a slope of 60.0° by a horizontal force. What is the normal force exerted by the book?

a. 39 N
b. 61 N
c. 15 N
d. 34 N

ANS: A   DIF: IIIA   OBJ: 4-4.2

40. A couch with a mass of 1×10^2 kg is placed on an adjustable ramp connected to a truck. As one end of the ramp is raised, the couch begins to move downward. If the couch slides down the ramp with an acceleration of 0.70 m/s² when the ramp angle is 25°, what is the coefficient of kinetic friction between the ramp and couch?
($g = 9.81$ m/s²)

a. 0.47
b. 0.42
c. 0.39
d. 0.12

ANS: C   DIF: IIIC   OBJ: 4-4.4

41. There are six books in a stack, and each book weighs 5 N. The coefficient of friction between the books is 0.2. With what horizontal force must one push to start sliding the top five books off the bottom one?

a. 1 N
b. 5 N
c. 3 N
d. 7 N

ANS: B   DIF: IIIB   OBJ: 4-4.4

42. A crate is carried in a pickup truck traveling horizontally at 15.0 m/s. The truck applies the brakes for a distance of 28.7 m while stopping with uniform acceleration. What is the coefficient of static friction between the crate and the truck bed if the crate does not slide?

a. 0.400
b. 0.365
c. 0.892
d. 0.656

ANS: A   DIF: IIIB   OBJ: 4-4.4

43. An ice skater moving at 10.0 m/s coasts to a halt in 1.0×10^2 m on a smooth ice surface. What is the coefficient of friction between the ice and the skates?
   a. 0.025
   b. 0.051
   c. 0.102
   d. 0.205

   ANS: B   DIF: IIIB   OBJ: 4-4.4

44. An Olympic skier moving at 20.0 m/s down a 30.0° slope encounters a region of wet snow and slides 145 m before coming to a halt. What is the coefficient of friction between the skis and the snow?
   a. 0.540
   b. 0.740
   c. 0.116
   d. 0.470

   ANS: B   DIF: IIIC   OBJ: 4-4.4

SHORT ANSWER

1. Briefly describe how applying the brakes to stop a bicycle is an example of force.

    ANS:
    Force is the cause of an acceleration or the change in an object's velocity. Applying the brakes decelerates the bicycle (accelerates it in the negative direction) and causes a change in the bicycle's velocity as the bicycle slows down.

    DIF: I   OBJ: 4-1.1

2. How does the theory of field forces explain how objects could exert force on each other without touching?

    ANS:
    Objects exert forces on one another when their fields interact.

    DIF: I   OBJ: 4-1.2

3. Construct a free-body diagram of a car being towed.

    ANS:

    DIF: II   OBJ: 4-1.3

Holt Physics Assessment Item Listing
37

4. State Newton's first law of motion.

ANS:
An object at rest remains at rest and an object in motion continues in motion with constant velocity unless it experiences a net external force.

DIF: I   OBJ: 4-2.1

5. What happens to an object in motion when it experiences a nonzero net external force?

ANS:
The object experiences a change in motion.

DIF: I   OBJ: 4-2.1

6. What is the natural tendency of an object that is in motion?

ANS:
The natural condition for a moving object is to remain in motion once it has been set in motion.

DIF: I   OBJ: 4-2.1

7. Describe the forces acting on a car as it moves along a level highway in still air at a constant speed.

ANS:
Gravity exerts a downward force on the car that is balanced by the normal force of the road acting upward on the car. The car's forward motion is opposed by the friction between the road and the tires and by the resistance of the air. The sum of these opposing forces is balanced by an equal and opposite force exerted by the engine and applied to the tires, where the road exerts a reaction force that is directed forward.

DIF: II   OBJ: 4-2.2

8. If two teams playing tug-of-war pull on a rope with equal but opposite forces, what is the net external force on the rope?

ANS:
0.0 N

DIF: I   OBJ: 4-2.2

9. In the equation form of Newton's law, $\Sigma \mathbf{F} = m\mathbf{a}$, what does $\Sigma \mathbf{F}$ represent?

ANS:
$\Sigma \mathbf{F}$ is the vector sum of external forces acting on the object.

DIF: I   OBJ: 4-3.1

10. Why does it require much less force to accelerate a low-mass object than it does to accelerate a high-mass object the same amount?

   ANS:
   An object with smaller mass has less inertia, or tendency to maintain velocity, than does an object with greater mass.

   DIF: II   OBJ: 4-3.1

11. A block of wood supported by two concrete blocks is chopped in half by a karate instructor. Identify an action-reaction pair, and compare the forces exerted by each object.

   ANS:
   The hand exerts a force on the wood, and the wood exerts an equal force on the hand.

   DIF: I   OBJ: 4-3.3

12. Do action-reaction pairs result in equilibrium? Explain.

   ANS:
   No, action-reaction pairs do not result in equilibrium because they act on different objects.

   DIF: II   OBJ: 4-3.4

13. When a horse pulls on a cart, the cart pulls on the horse with an equal but opposite force. How is the horse able to pull the cart?

   ANS:
   The horse and the cart are not necessarily at equilibrium. The forces in the action-reaction pair are each applied to different objects.

   DIF: II   OBJ: 4-3.4

14. Distinguish between mass and weight.

   ANS:
   Mass is the amount of matter in an object and is an inherent property of an object. Weight is not an inherent property of an object and is the magnitude of the force of gravity acting on the object.

   DIF: II   OBJ: 4-4.1

15. In what direction does the force of air resistance act?

   ANS:
   Air resistance acts in the direction opposite the direction of an object's motion.

   DIF: I   OBJ: 4-4.3

16. What happens to air resistance when an object accelerates?

ANS:
In most cases, air resistance increases with increasing speed.

DIF: I   OBJ: 4-4.3

17. When a car is moving, what happens to the velocity and acceleration of the car if the air resistance becomes equal to the force acting in the opposite direction?

ANS:
The acceleration is then zero, and the car moves at a constant speed.

DIF: I   OBJ: 4-4.3

18. Why is air resistance considered a form of friction?

ANS:
Air resistance is a form of friction because it is a retarding force. It acts in the direction opposite an object's motion.

DIF: I   OBJ: 4-4.3

PROBLEM

1. In a game of tug-of-war, a rope is pulled to the left with a force of 75 N and to the right with a force of 102 N. What is the magnitude and direction of the net external force on the rope?

ANS:
27 N to the right

DIF: IIIA   OBJ: 4-2.2

MULTIPLE CHOICE

1. A force does work on an object if a component of the force
    a. is perpendicular to the displacement of the object.
    b. is parallel to the displacement of the object.
    c. perpendicular to the displacement of the object moves the object along a path that returns the object to its starting position.
    d. parallel to the displacement of the object moves the object along a path that returns the object to its starting position.

    ANS: B   DIF: I   OBJ: 5-1.2

2. What is the common formula for work?
    a. $W = F\Delta x$ c. $W = Fd^2$
    b. $W = Fd$ d. $W = F^2d$

    ANS: A   DIF: I   OBJ: 5-1.2

3. Work is done when
    a. the displacement is not zero.
    b. the displacement is zero.
    c. the force is zero.
    d. the force and displacement are perpendicular.

    ANS: A   DIF: I   OBJ: 5-1.2

4. A 1.00×10^3 kg sports car accelerates from rest to 25.0 m/s in 7.50 s. What is the average power output of the automobile engine?
    a. 20.8 kW   c. 41.7 kW
    b. 30.3 kW   d. 52.4 kW

    ANS: C   DIF: IIIB   OBJ: 5-4.3

5. The more powerful the motor is,
    a. the longer the time interval for doing the work is.
    b. the shorter the time interval for doing the work is.
    c. the greater the ability to do the work is.
    d. the shorter the workload is.

    ANS: B   DIF: I   OBJ: 5-4.4

6. The magnitude of the component of the force that does the work is 43.0 N. How much work is done on a bookshelf being pulled 5.00 m at an angle of 37.0° from the horizontal?
    a. 172 J   c. 129 J
    b. 215 J   d. 792 J

    ANS: B   DIF: IIIB   OBJ: 5-1.4

7. A worker pushes a wheelbarrow with a horizontal force of 50.0 N over a level distance of 5.0 m. If a frictional force of 43 N acts on the wheelbarrow in a direction opposite to that of the worker, what net work is done on the wheelbarrow?
    a. 250 J
    b. 0.0 J
    c. 35 J
    d. 10.0 J

    ANS: C   DIF: IIIA   OBJ: 5-1.4

8. A hill is 100 m long and makes an angle of 12° with the horizontal. As a 50 kg jogger runs up the hill, how much work does gravity do on the jogger?
    a. 50 000 J
    b. 10 000 J
    c. −10 000 J
    d. 0.0 J

    ANS: B   DIF: IIIA   OBJ: 5-1.4

9. A child moving at constant velocity carries a 2 N ice-cream cone 1 m across a level surface. What is the net work done on the ice-cream cone?
    a. 0 J
    b. 0.5 J
    c. 2 J
    d. 20 J

    ANS: A   DIF: IIIA   OBJ: 5-1.4

10. A construction worker pushes a wheelbarrow 5.0 m with a horizontal force of 50.0 N. How much work is done by the worker on the wheelbarrow?
    a. 10 J
    b. 1250 J
    c. 250 J
    d. 55 J

    ANS: C   DIF: IIIA   OBJ: 5-1.4

11. A horizontal force of 200 N is applied to move a 55 kg television set across a 10 m level surface. What is the work done by the 200 N force on the television set?
    a. 4000 J
    b. 5000 J
    c. 2000 J
    d. 6000 J

    ANS: C   DIF: IIIA   OBJ: 5-1.4

12. A flight attendant pulls a 50.0 N flight bag a distance of 250.0 m along a level airport floor at a constant speed. A 30.0 N force is exerted on the bag at an angle of 50.0° above the horizontal. How much work is done on the flight bag?
    a. 12 500 J
    b. 7510 J
    c. 4820 J
    d. 8040 J

    ANS: C   DIF: IIIB   OBJ: 5-1.4

13. Which of the following energy forms is the sum of kinetic energy and all forms of potential energy?
    a. total energy
    b. sum (Σ) energy
    c. nonmechanical energy
    d. mechanical energy

    ANS: D   DIF: I   OBJ: 5-2.1

14. Which of the following energy forms is involved in winding a pocket watch?
    a. electrical energy
    b. nonmechanical energy
    c. gravitational potential energy
    d. elastic potential energy

    ANS: D   DIF: I   OBJ: 5-2.1

15. Which of the following energy forms is NOT involved in hitting a tennis ball?
    a. kinetic energy
    b. chemical potential energy
    c. gravitational potential energy
    d. elastic potential energy

    ANS: A   DIF: I   OBJ: 5-2.1

16. Which of the following energy forms is involved in a pencil falling from a desk?
    a. kinetic energy
    b. nonmechanical energy
    c. gravitational potential energy
    d. elastic potential energy and kinetic energy

    ANS: C   DIF: I   OBJ: 5-2.1

17. A 3.00 kg toy falls from a height of 10.0 m. Just before hitting the ground, what will be its kinetic energy? (Disregard air resistance. $g = 9.81$ m/s^2.)
    a. 98.0 J
    b. 0.98 J
    c. 29.4 J
    d. 294 J

    ANS: D   DIF: IIIA   OBJ: 5-2.2

18. If the only force acting on an object is friction during a given physical process, which of the following assumptions must be made in regard to the object's kinetic energy?
    a. The kinetic energy decreases.
    b. The kinetic energy increases.
    c. The kinetic energy remains constant.
    d. The kinetic energy decreases and then increases.

    ANS: A   DIF: I   OBJ: 5-2.2

19. What is the kinetic energy of a 0.135 kg baseball thrown at 40.0 m/s?
    a. 54.0 J
    b. 87.0 J
    c. 108 J
    d. 216 J

    ANS: C   DIF: IIIA   OBJ: 5-2.2

20. If both the mass and the velocity of a ball are tripled, the kinetic energy of the ball is increased by a factor of
    a. 3.
    b. 6.
    c. 9.
    d. 27.

    ANS: D   DIF: II   OBJ: 5-2.2

21. Which of the following energy forms is associated with an object in motion?
    a. potential energy
    b. elastic potential energy
    c. nonmechanical energy
    d. kinetic energy

    ANS: D   DIF: I   OBJ: 5-2.3

22. Which of the following energy forms is associated with an object due to its position?
    a. potential
    b. positional
    c. total
    d. kinetic

    ANS: A   DIF: I   OBJ: 5-2.3

23. The main difference between kinetic energy and potential energy is that
    a. kinetic energy involves position and potential energy involves motion.
    b. kinetic energy involves motion and potential energy involves position.
    c. although both energies involve motion, only kinetic involves position.
    d. although both energies involve position, only potential involves motion.

    ANS: B   DIF: I   OBJ: 5-2.3

24. Which of the following energy forms is associated with an object due to its position relative to Earth?
    a. potential energy
    b. elastic potential energy
    c. gravitational potential energy
    d. kinetic energy

    ANS: C   DIF: I   OBJ: 5-2.4

25. Which of the following energy forms is stored in any compressed or stretched object?
    a. nonmechanical energy
    b. elastic potential energy
    c. gravitational potential energy
    d. kinetic energy

    ANS: B   DIF: I   OBJ: 5-2.4

26. The equation for determining gravitational potential energy is $PE_g = mgh$. Which factor(s) in this equation is (are) NOT a property of an object?
    a. g
    b. h
    c. m
    d. both g and h

    ANS: D   DIF: I   OBJ: 5-2.4

27. Which of the following parameters does not express how resistant a spring is to being compressed or stretched?
    a. compression distance
    b. relaxed length
    c. spring constant
    d. stretching distance

    ANS: C   DIF: I   OBJ: 5-2.4

28. Which form of energy is involved in weighing fruit on a spring scale?
    a. kinetic energy
    b. nonmechanical energy
    c. gravitational potential energy
    d. elastic potential energy

    ANS: D   DIF: I   OBJ: 5-2.4

29. Which of the following energy forms is associated with an object's interaction with the environment?
    a. potential energy
    b. kinetic energy
    c. mechanical energy
    d. nonmechanical energy

    ANS: A   DIF: I   OBJ: 5-2.5

30. As an object is lowered into a deep hole in the ground, which of the following assumptions must be made in regard to the object's potential energy?
    a. The potential energy increases.
    b. The potential energy decreases.
    c. The potential energy remains constant.
    d. The potential energy increases and then decreases.

    ANS: B   DIF: I   OBJ: 5-2.5

31. A 40.0 N crate is pulled up a 5.0 m inclined plane at a constant velocity. If the plane is inclined at an angle of 37° to the horizontal and there is a constant force of friction of 10.0 N between the crate and the surface, what is the net gain in potential energy by the crate?
    a. 120 J   c. 210 J
    b. −120 J   d. −210 J

    ANS: A   DIF: IIIB   OBJ: 5-2.5

32. A 0.002 kg coin, which has zero potential energy at rest, is dropped into a 10.0 m well. After the coin comes to a stop in the mud, what is its potential energy?
    a. 0.000 J   c. −0.196 J
    b. 0.196 J   d. 0.020 J

    ANS: C   DIF: IIIB   OBJ: 5-2.5

33. An 80.0 kg climber with a 20.0 kg pack climbs 8848 m to the top of Mount Everest. What is the climber's potential energy?
    a. 6.94×10^6 J   c. 2.47×10^6 J
    b. 4.16×10^6 J   d. 1.00×10^6 J

    ANS: A   DIF: IIIB   OBJ: 5-2.5

34. A 5.00×10^2 N crate is at the top of a 5.00 m ramp, which is inclined at 20.0° with the horizontal. What is its potential energy? ($g = 9.81$ m/s².)
    a. 855 J   c. 815 J
    b. 2350 J   d. 8390 J

    ANS: A   DIF: IIIB   OBJ: 5-2.5

35. In the presence of frictional force,
    a. nonmechanical energy is negligible and mechanical energy is no longer conserved.
    b. nonmechanical energy is negligible and mechanical energy is conserved.
    c. nonmechanical energy is no longer negligible and mechanical energy is conserved.
    d. nonmechanical energy is no longer negligible and mechanical energy is no longer conserved.

    ANS: D   DIF: I

36. Why doesn't the principle of mechanical energy conservation hold in situations when frictional forces are present?
    a. Kinetic energy is not simply converted to a form of potential energy.
    b. Potential energy is simply converted to a form of gravitational energy.
    c. Chemical energy is not simply converted to electrical energy.
    d. Kinetic energy is simply converted to a form of gravitational energy.

    ANS: A   DIF: I   OBJ: 5-3.2

37. Which of the following are examples of conservable quantities?
    a. potential energy and length   c. mechanical energy and mass
    b. mechanical energy and length   d. kinetic energy and mass

    ANS: C   DIF: I   OBJ: 5-3.2

38. A 16.0 kg child on roller skates, initially at rest, rolls 2.0 m down an incline at an angle of 20.0° with the horizontal. If there is no friction between incline and skates, what is the kinetic energy of the child at the bottom of the incline? ($g = 9.81$ m/s^2.)
    a. 210 J   c. 11 J
    b. 610 J   d. 110 J

    ANS: D   DIF: IIIA   OBJ: 5-3.3

39. Old Faithful geyser in Yellowstone National Park shoots water every hour to a height of 40.0 m. With what velocity does the water leave the ground? (Disregard air resistance. $g = 9.81$ m/s^2.)
    a. 7.00 m/s   c. 19.8 m/s
    b. 14.0 m/s   d. 28.0 m/s

    ANS: D   DIF: IIIB   OBJ: 5-3.3

40. A pole vaulter clears 6.00 m. With what velocity does the vaulter strike the mat in the landing area? (Disregard air resistance. $g = 9.81$ m/s^2.)
    a. 2.70 m/s   c. 10.8 m/s
    b. 5.40 m/s   d. 21.6 m/s

    ANS: C   DIF: IIIB   OBJ: 5-3.3

41. A bobsled zips down an ice track starting at 150 m vertical distance up the hill. Disregarding friction, what is the velocity of the bobsled at the bottom of the hill? ($g = 9.81$ m/s^2.)
    a. 27 m/s   c. 45 m/s
    b. 36 m/s   d. 54 m/s

    ANS: D   DIF: IIIB   OBJ: 5-3.3

42. A professional skier starts from rest and reaches a speed of 56 m/s on a ski slope 30.0° above the horizontal. Using the work–kinetic energy theorem and disregarding friction, find the minimum distance along the slope the skier would have to travel in order to reach this speed.
    a. 110 m   c. 320 m
    b. 160 m   d. 640 m

    ANS: C   DIF: IIIB   OBJ: 5-4.1

43. A 40.0 N crate starting at rest slides down a rough 6.0 m long ramp inclined at 30.0° with the horizontal. The force of friction between the crate and ramp is 6.0 N. Using the work–kinetic energy theorem, find the velocity of the crate at the bottom of the incline.
   a. 8.7 m/s
   b. 3.3 m/s
   c. 4.5 m/s
   d. 6.4 m/s

   ANS: D   DIF: IIIB   OBJ: 5-4.1

44. A 15.0 kg crate, initially at rest, slides down a ramp 2.0 m long and inclined at an angle of 20.0° with the horizontal. Using the work–kinetic energy theorem and disregarding friction, find the velocity of the crate at the bottom of the ramp. ($g = 9.81$ m/s^2.)
   a. 6.1 m/s
   b. 3.7 m/s
   c. 9.7 m/s
   d. 8.3 m/s

   ANS: B   DIF: IIIB   OBJ: 5-4.1

45. A parachutist with a mass of 50.0 kg jumps out of an airplane at an altitude of 1.00×10^3 m. After the parachute deploys, the parachutist lands with a velocity of 5.00 m/s. Using the work–kinetic energy theorem, find the energy that was lost to air resistance during this jump. ($g = 9.81$ m/s^2.)
   a. 49 300 J
   b. 98 800 J
   c. 198 000 J
   d. 489 000 J

   ANS: D   DIF: IIIB   OBJ: 5-4.1

46. A horizontal force of 2.00×10^2 N is applied to a 55.0 kg cart across a 10.0 m level surface, accelerating it 2.00 m/s^2. Using the work–kinetic energy theorem, find the force of friction that slows the motion of the cart? (Disregard air resistance. $g = 9.81$ m/s^2.)
   a. 110 N
   b. 90.0 N
   c. 80.0 N
   d. 70.0 N

   ANS: B   DIF: IIIB   OBJ: 5-4.1

47. A child riding a bicycle has a total mass of 40.0 kg. The child approaches the top of a hill that is 10.0 m high and 100.0 m long at 5.0 m/s. If the force of friction between the bicycle and the hill is 20.0 N, what is the child's velocity at the bottom of the hill? (Disregard air resistance. $g = 9.81$ m/s^2.)
   a. 5.0 m/s
   b. 10.0 m/s
   c. 11 m/s
   d. The child stops before reaching the bottom.

   ANS: C   DIF: IIIB   OBJ: 5-4.1

48. Which of the following is the rate at which energy is transferred?
   a. potential energy
   b. kinetic energy
   c. mechanical energy
   d. power

   ANS: D   DIF: I   OBJ: 5-4.2

49. Which of the following equations is NOT an equation for power?

a. $P = F\dfrac{d}{\Delta t}$

c. $P = Fv$

b. $P = \dfrac{W}{\Delta t}$

d. $P = \dfrac{Fv}{\Delta t}$

ANS: D   DIF: I   OBJ: 5-4.2

50. What is the average power supplied by a 60.0 kg secretary running up a flight of stairs rising vertically 4.0 m in 4.2 s?

a. 380 W
b. 560 W

c. 610 W
d. 670 W

ANS: B   DIF: IIIB   OBJ: 5-4.3

51. What is the average power output of a weight lifter who can lift 250 kg 2.0 m in 2.0 s?

a. 5.0×10^2 W
b. 2.5 kW

c. 4.9 kW
d. 9.8 kW

ANS: B   DIF: IIIB   OBJ: 5-4.3

52. A jet engine develops 1.0×10^5 N of thrust to move an airplane forward at a speed of 9.0×10^2 km/h. What is the power output of the engine?

a. 550 kW
b. 1.0 MW

c. 25 MW
d. 5.0 MW

ANS: C   OBJ: 5-4.3

53. Water flows over a section of Niagara Falls at a rate of 1.20×10^6 kg/s and falls 50.0 m. What is the power of the waterfall?

a. 589 MW
b. 294 MW

c. 147 MW
d. 60.0 MW

ANS: A   DIF: IIIC   OBJ: 5-4.3

SHORT ANSWER

1. In the following sentence, is the everyday or the scientific meaning of work intended?
A student works on a term paper.

ANS:
everyday meaning

DIF: I   OBJ: 5-1.1

2. In the following sentence, is the everyday or the scientific meaning of work intended?
   A coach does work on the bleachers by moving them into place before the basketball game.

   ANS:
   scientific meaning

   DIF: I   OBJ: 5-1.1

3. In the following sentence, is the everyday or the scientific meaning of work intended?
   A road cleaner does work shoveling snow.

   ANS:
   scientific meaning

   DIF: I   OBJ: 5-1.1

4. In the following sentence, is the everyday or the scientific meaning of work intended?
   A clerk works overtime on Saturdays.

   ANS:
   everyday meaning

   DIF: I   OBJ: 5-1.1

5. How is work related to force and displacement?

   ANS:
   Work is equal to the magnitude of the component of a force parallel to the displacement of the
   object multiplied by the displacement of an object.

   DIF: I   OBJ: 5-1.2

6. A child pulls a toy across the floor. Is the work done on the toy positive or negative?

   ANS:
   positive

   DIF: I   OBJ: 5-1.3

7. A worker picks up a bucket and sets it back down again in the same place. Is net work done on
   the bucket?

   ANS:
   no

   DIF: I   OBJ: 5-1.3

8. Air exerts a force on a leaf as it falls from a tree to Earth. Is the work done on the leaf positive or negative?

ANS:
negative

DIF: I   OBJ: 5-1.3

9. Is conservation of mechanical energy likely to hold in this situation?
A soccer ball is thrown into the air. (Disregard air resistance.)

ANS:
yes

DIF: I   OBJ: 5-3.1

10. Is conservation of mechanical energy likely to hold in this situation?
A crystal vase falls off the table. (Disregard air resistance.)

ANS:
yes

DIF: I   OBJ: 5-3.1

11. Is conservation of mechanical energy likely to hold in this situation?
A skateboard rolls across a sewer grate.

ANS:
no

DIF: I   OBJ: 5-3.1

12. Is conservation of mechanical energy likely to hold in this situation?
An ice skater glides across freshly made ice.

ANS:
no

DIF: I   OBJ: 5-3.1

13. Explain how energy, time, and power are related.

ANS:
Power is the rate at which energy is transferred. In other words, power is the energy transferred in a given time interval.

DIF: I   OBJ: 5-4.2

14. What does the wattage of a light bulb indicate?

ANS:
The wattage tells the rate at which energy is converted by the bulb.

DIF: II   OBJ: 5-4.2

15. Which motor performs more work in the same amount of time—a 10 kW motor or a 20 kW motor? Explain your reasoning.

ANS:
The more powerful 20 kW motor would do work in a shorter amount of time (or the 20 kW motor does twice as much work in the same amount of time).

DIF: I   OBJ: 5-4.4

16. What occurs when a machine does work on an object?

ANS:
When a machine does work on an object, energy is transferred to that object.

DIF: I   OBJ: 5-4.4

17. How is a machine's power rating related to its rate of doing work on an object?

ANS:
The power rating of a machine indicates the rate at which it does work on an object. A machine with a high power rating can perform the same amount of work as a lower-rated machine in less time—that is its rate.

DIF: I   OBJ: 5-4.4

PROBLEM

1. A skier with a mass of 88 kg hits a ramp of snow at 16 m/s and becomes airborne. At the highest point of flight, the skier is 3.7 m above the ground. What is the skier's gravitational potential energy at this point?

ANS:
3.2×10^3 J

DIF: IIIA   OBJ: 5-2.5

MULTIPLE CHOICE

1. Which of the following has the greatest momentum?
    a. truck with a mass of 2250 kg moving at a velocity of 25 m/s
    b. car with a mass of 1210 kg moving at a velocity of 51 m/s
    c. truck with a mass of 6120 kg moving at a velocity of 10 m/s
    d. car with a mass of 1540 kg moving at a velocity of 38 m/s

    ANS: B   DIF: IIIB   OBJ: 6-1.1

2. Which of the following has the greatest momentum?
    a. tortoise with a mass of 270 kg moving at a velocity of 0.5 m/s
    b. hare with a mass of 2.7 kg moving at a velocity of 7 m/s
    c. turtle with a mass of 91 kg moving at a velocity of 1.4 m/s
    d. roadrunner with a mass of 1.8 kg moving at a velocity of 6.7 m/s

    ANS: A   DIF: IIIB   OBJ: 6-1.1

3. What velocity must a 1340 kg car have in order to have the same momentum as a 2680 kg truck traveling at a velocity of 15 m/s to the west?
    a. 6.0×10^1 m/s to the west
    c. 3.0×10^1 m/s to the west
    b. 6.0×10^1 m/s to the east
    d. 3.0×10^1 m/s to the east

    ANS: C   DIF: IIIB   OBJ: 6-1.1

4. A child with a mass of 23 kg rides a bike with a mass of 5.5 kg at a velocity of 4.5 m/s to the south. Compare the momentum of the child with the momentum of the bike.
    a. Both the child and the bike have the same momentum.
    b. The bike has a greater momentum than the child.
    c. The child has a greater momentum than the bike.
    d. Neither the child nor the bike has momentum.

    ANS: C   DIF: II   OBJ: 6-1.1

5. When comparing the momentum of two moving objects, which of the following is correct?
    a. The object with the higher velocity will have less momentum if the masses are equal.
    b. The more massive object will have less momentum if its velocity is greater.
    c. The less massive object will have less momentum if the velocities are the same.
    d. The more massive object will have less momentum if the velocities are the same.

    ANS: C   DIF: II   OBJ: 6-1.1

6. A baseball is pitched very fast. Another baseball of equal mass is pitched very slowly. Which of the following statements is correct?
    a. The fast-moving baseball is harder to stop because it has more momentum.
    b. The slow-moving baseball is harder to stop because it has more momentum.
    c. The fast-moving baseball is easier to stop because it has more momentum.
    d. The slow-moving baseball is easier to stop because it has more momentum.

    ANS: A   DIF: II   OBJ: 6-1.2

7. A roller coaster climbs up a hill at 4 m/s and then zips down the hill at 30 m/s. The momentum of the roller coaster
   a. is greater up the hill than down the hill.   c. remains the same throughout the ride.
   b. is greater down the hill than up the hill.   d. is zero throughout the ride.

   ANS: B   DIF: II   OBJ: 6-1.2

8. A person sitting in a chair with wheels stands, causing the chair to roll backward across the floor. The momentum of the chair
   a. was zero while stationary and increased when the person stood.
   b. was greatest while the person sat in the chair.
   c. remained the same.
   d. was zero when the person got out of the chair and increased while the person sat.

   ANS: A   DIF: II   OBJ: 6-1.2

9. A student walks to class at a velocity of 3 m/s. To avoid walking into a door as it opens, the student slows to a velocity of 0.5 m/s. Now late for class, the student runs down the corridor at a velocity of 7 m/s. The student had the least momentum
   a. while walking at a velocity of 3 m/s.
   b. while dodging the opening door.
   c. immediately after the door opened.
   d. while running to class at a velocity of 7 m/s.

   ANS: B   DIF: II   OBJ: 6-1.2

10. An ice skater initially skating at a velocity of 3 m/s speeds up to a velocity of 5 m/s. The momentum of the skater
    a. decreases.   c. remains the same.
    b. increases.   d. becomes zero.

    ANS: B   DIF: II   OBJ: 6-1.2

11. If a force is exerted on an object, which statement is true?
    a. A large force always produces a large change in the object's momentum.
    b. A large force produces a large change in the object's momentum only if the force is applied over a very short time interval.
    c. A small force applied over a long time interval can produce a large change in the object's momentum.
    d. A small force produces a large change in the object's momentum.

    ANS: C   DIF: II   OBJ: 6-1.3

12. The change in an object's momentum is equal to
    a. the product of the mass of the object and the time interval.
    b. the product of the force applied to the object and the time interval.
    c. the time interval divided by the net external force.
    d. the net external force divided by the time interval.

    ANS: B   DIF: I   OBJ: 6-1.3

13. A force is applied to stop a moving shopping cart. Increasing the time interval over which the force is applied
    a. requires a greater force.   c. requires a smaller force.
    b. has no effect on the force needed.   d. requires the same force.

    ANS: C   DIF: II   OBJ: 6-1.3

14. Which of the following situations is an example of a visible change in momentum?
    a. A hiker walks through a spider's web.   c. A volleyball hits a mosquito in the air.
    b. A car drives over a pebble.   d. A baseball is hit by a bat.

    ANS: D   DIF: II   OBJ: 6-1.3

15. Which of the following situations is an example of change in momentum?
    a. A tennis ball is hit into a net.
    b. A helium-filled balloon rises upward into the sky.
    c. An airplane flies into some scattered white clouds.
    d. A bicyclist rides over a leaf on the pavement.

    ANS: A   DIF: II   OBJ: 6-1.3

16. A 6.0×10^{-2} kg tennis ball moves at a velocity of 12 m/s. The ball is struck by a racket, causing it to rebound in the opposite direction at a speed of 18 m/s. What is the change in the ball's momentum?
    a. −0.38 kg•m/s   c. −1.1 kg•m/s
    b. −0.72 kg•m/s   d. −1.8 kg•m/s

    ANS: D   DIF: IIIB   OBJ: 6-1.3

17. A rubber ball with a mass of 0.30 kg is dropped onto a steel plate. The ball's velocity just before impact is 4.5 m/s and just after impact is 4.2 m/s. What is the change in the ball's momentum?
    a. −0.09 kg•m/s   c. −4.0 kg•m/s
    b. −2.6 kg•m/s   d. −12 kg•m/s

    ANS: B   DIF: IIIB   OBJ: 6-1.3

18. A 0.2 baseball if pitched with a velocity of 40 m/s and is then batted to the pitcher with a velocity of 60 m/s. What is the magnitude of change in the ball's momentum?
    a. 4 kg•m/s   c. 2 kg•m/s
    b. 8 kg•m/s   d. 20 kg•m/s

    ANS: D   DIF: IIIB   OBJ: 6-1.3

19. A ball with a momentum of 4.0 kg•m/s hits a wall and bounces straight back without losing any kinetic energy. What is the change in the ball's momentum?
    a. 0.0 kg•m/s   c. 8.0 kg•m/s
    b. −4.0 kg•m/s   d. −8.0 kg•m/s

    ANS: D   DIF: IIIA   OBJ: 6-1.3

20. A softball with a mass of 0.11 kg moves at a speed of 12 m/s. Then the ball is hit by a bat and rebounds in the opposite direction at a speed of 15 m/s. What is the change in momentum of the ball?
    a. −1.3 kg•m/s
    b. −1.6 kg•m/s
    c. −0.33 kg•m/s
    d. −3.0 kg•m/s

    ANS: D   DIF: IIIB   OBJ: 6-1.3

21. A ball with a mass of 0.15 kg and a velocity of 5.0 m/s strikes a wall and bounces straight back with a velocity of 3.0 m/s. What is the change in momentum of the ball?
    a. −0.30 kg•m/s
    b. −1.20 kg•m/s
    c. −0.15 kg•m/s
    d. −7.50 kg•m/s

    ANS: B   DIF: IIIB   OBJ: 6-1.3

22. The impulse experienced by a body is equivalent to the body's change in
    a. velocity.
    b. kinetic energy.
    c. momentum.
    d. force.

    ANS: C   DIF: I   OBJ: 6-1.4

23. A moderate force will break an egg. However, an egg dropped on the road usually breaks, while one dropped on the grass usually does not break because for the egg dropped on the grass,
    a. the change in momentum is greater.
    b. the change in momentum is less.
    c. the time interval for stopping is greater.
    d. the time interval for stopping is less.

    ANS: C   DIF: I   OBJ: 6-1.4

24. Which of the following statements properly relates the variables in the equation $F\Delta t = \Delta p$?
    a. A large constant force changes an object's momentum over a long time interval.
    b. A large constant force acting over a long time interval causes a large change in momentum.
    c. A large constant force changes an object's momentum at various time intervals.
    d. A large constant force does not necessarily cause a change in an object's momentum.

    ANS: B   DIF: I   OBJ: 6-1.4

25. A large moving ball collides with a small stationary ball. The momentum
    a. of the large ball decreases, and the momentum of the small ball increases.
    b. of the small ball decreases, and the momentum of the large ball increases.
    c. of the large ball increases, and the momentum of the small ball decreases.
    d. does not change for either ball.

    ANS: A   DIF: II   OBJ: 6-2.1

26. A 75 kg person walking around a corner bumped into an 80 kg person who was running around the same corner. The momentum of the 80 kg person
    a. increased.
    b. decreased.
    c. remained the same.
    d. was conserved.

    ANS: B   DIF: II   OBJ: 6-2.1

27. A 20 kg shopping cart moving at a velocity of 0.5 m/s collides into a store wall and stops. The momentum of the shopping cart
   a. increases.
   b. decreases.
   c. remains the same.
   d. is conserved.

   ANS: B   DIF: II   OBJ: 6-2.1

28. A rubber ball moving at a speed of 5 m/s hit a flat wall and returned to the thrower at 5 m/s. The magnitude of the momentum of the rubber ball
   a. increased.
   b. decreased.
   c. remained the same.
   d. was not conserved.

   ANS: C   DIF: II   OBJ: 6-2.1

29. Two objects with different masses collide and bounce back after an elastic collision. Before the collision, the two objects were moving at velocities equal in magnitude but opposite in direction. After the collision,
   a. the less massive object had gained momentum.
   b. the more massive object had gained momentum.
   c. both objects had the same momentum.
   d. both objects lost momentum.

   ANS: A   DIF: II   OBJ: 6-2.1

30. Two skaters stand facing each other. One skater's mass is 60 kg, and the other's mass is 72 kg. If the skaters push away from each other without spinning,
   a. the 60 kg skater travels at a lower momentum.
   b. their momenta are equal but opposite.
   c. their total momentum doubles.
   d. their total momentum decreases.

   ANS: B   DIF: II   OBJ: 6-2.2

31. Two swimmers relax close together on air mattresses in a pool. One swimmer's mass is 48 kg, and the other's mass is 55 kg. If the swimmers push away from each other,
   a. their total momentum triples.
   b. their momenta are equal but opposite.
   c. their total momentum doubles.
   d. their total momentum decreases.

   ANS: B   DIF: II   OBJ: 6-2.2

32. A soccer ball collides with another soccer ball at rest. The total momentum of the balls
   a. is zero.
   b. increases.
   c. remains constant.
   d. decreases.

   ANS: C   DIF: II   OBJ: 6-2.2

33. Paint is splattered on a canvas. After the paint sticks to the canvas, the total momentum of the paint and canvas
   a. is zero.
   b. increases.
   c. is equal and opposite.
   d. decreases.

   ANS: A   DIF: II   OBJ: 6-2.2

34. In a two-body collision,
    a. momentum is conserved.
    b. kinetic energy is conserved.
    c. neither momentum nor kinetic energy is conserved.
    d. both momentum and kinetic energy are conserved.

    ANS: A   DIF: I   OBJ: 6-2.3

35. The law of conservation of momentum states that
    a. the total initial momentum of all objects interacting with one another usually equals the total final momentum.
    b. the total initial momentum of all objects interacting with one another does not equal the total final momentum.
    c. the total momentum of all objects interacting with one another is zero.
    d. the total momentum of all objects interacting with one another remains constant regardless of the nature of the forces between the objects.

    ANS: D   DIF: I   OBJ: 6-2.3

36. Which of the following statements about the conservation of momentum is NOT correct?
    a. Momentum is conserved for a system of objects pushing away from each other.
    b. Momentum is not conserved for a system of objects in a head-on collision.
    c. Momentum is conserved when two or more interacting objects push away from each other.
    d. The total momentum of a system of interacting objects remains constant regardless of forces between the objects.

    ANS: B   DIF: I   OBJ: 6-2.3

37. A swimmer with a mass of 75 kg dives off a raft with a mass of 500 kg. If the swimmer's speed is 4 m/s immediately after leaving the raft, what is the speed of the raft?
    a. 0.2 m/s   c. 0.6 m/s
    b. 0.5 m/s   d. 4.0 m/s

    ANS: C   DIF: IIIB   OBJ: 6-2.4

38. An astronaut with a mass of 70.0 kg is outside a space capsule when the tether line breaks. To return to the capsule, the astronaut throws a 2.0 kg wrench away from the capsule at a speed of 14 m/s. At what speed does the astronaut move toward the capsule?
    a. 5.0 m/s   c. 3.5 m/s
    b. 0.4 m/s   d. 7.0 m/s

    ANS: B   DIF: IIIB   OBJ: 6-2.4

39. A bullet with a mass of 5.00×10^{-3} kg is loaded into a gun. The loaded gun has a mass of 0.52 kg. The bullet is fired, causing the empty gun to recoil at a speed of 2.1 m/s. What is the speed of the bullet?
    a. 48 m/s   c. 120 m/s
    b. 220 m/s   d. 360 m/s

    ANS: B   DIF: IIIC   OBJ: 6-2.4

40. A 65.0 kg ice skater standing on frictionless ice throws a 0.15 kg snowball horizontally at a speed of 32.0 m/s. At what velocity does the skater move backward?
    a. 0.07 m/s
    b. 0.30 m/s
    c. 0.15 m/s
    d. 1.20 m/s

    ANS: A   DIF: IIIB   OBJ: 6-2.4

41. Two skaters, each with a mass of 50 kg, are stationary on a frictionless ice pond. One skater throws a 0.2 kg ball at 5 m/s to the other skater, who catches it. What are the velocities of the skaters when the ball is caught?
    a. 0.02 m/s moving apart
    b. 0.04 m/s moving apart
    c. 0.02 m/s moving toward each other
    d. 0.04 m/s moving toward each other

    ANS: A   DIF: IIIB   OBJ: 6-2.4

42. Two carts with masses of 1.5 kg and 0.7 kg, respectively, are held together by a compressed spring. When released, the 1.5 kg cart moves to the left with a velocity of 7 m/s. What is the velocity of the 0.7 kg cart? (Disregard the mass of the spring.)
    a. 15 m/s to the right
    b. 15 m/s to the left
    c. 7 m/s to the right
    d. 7 m/s to the left

    ANS: A   DIF: IIIA   OBJ: 6-2.4

43. Each croquet ball in a set has a mass of 0.50 kg. The green ball travels at 10.5 m/s and strikes a stationary red ball. If the green ball stops moving, what is the final speed of the red ball after the collision?
    a. 10.5 m/s
    b. 6.0 m/s
    c. 12.0 m/s
    d. 9.6 m/s

    ANS: A   DIF: IIIB   OBJ: 6-2.4

44. A diver with a mass of 80.0 kg jumps from a dock into a 130.0 kg boat at rest on the west side of the dock. If the velocity of the diver in the air is 4.10 m/s to the west, what is the final velocity of the diver after landing in the boat?
    a. 2.52 m/s to the west
    b. 2.52 m/s to the east
    c. 1.56 m/s to the west
    d. 1.56 m/s to the east

    ANS: C   DIF: IIIB   OBJ: 6-2.4

45. Two objects move separately after colliding, and both the total momentum and total kinetic energy remain constant. Identify the type of collision.
    a. elastic
    b. perfectly elastic
    c. inelastic
    d. perfectly inelastic

    ANS: A   DIF: I   OBJ: 6-3.1

46. Two objects stick together and move with the same velocity after colliding. Identify the type of collision.
    a. elastic
    b. perfectly elastic
    c. inelastic
    d. perfectly inelastic

    ANS: D   DIF: I   OBJ: 6-3.1

47. After colliding, objects are deformed and lose some kinetic energy. Identify the type of collision.
    a. elastic
    c. inelastic
    b. perfectly elastic
    d. perfectly inelastic

    ANS: C   DIF: I   OBJ: 6-3.1

48. Two billiard balls collide. Identify the type of collision.
    a. elastic
    c. inelastic
    b. perfectly elastic
    d. perfectly inelastic

    ANS: A   DIF: I   OBJ: 6-3.1

49. Two balls of dough collide and stick together. Identify the type of collision.
    a. elastic
    c. inelastic
    b. perfectly elastic
    d. perfectly inelastic

    ANS: D   DIF: I   OBJ: 6-3.1

50. Two snowballs with masses of 0.40 kg and 0.60 kg, respectively, collide head-on and combine to form a single snowball. The initial speed for each is 15 m/s. If the velocity of the snowball with a mass of 1.0 kg is 3.0 m/s after the collision, what is the decrease in kinetic energy?
    a. zero
    c. 60 J
    b. 110 J
    d. 90 J

    ANS: B   DIF: IIIB   OBJ: 6-3.2

51. A 1.5×10^3 kg truck moving at 15 m/s strikes a 7.5×10^2 kg automobile stopped at a traffic light. The vehicles hook bumpers and skid together at 10.0 m/s. What is the decrease in kinetic energy?
    a. 1.1×10^5 J
    c. 1.7×10^5 J
    b. 1.2×10^4 J
    d. 6.0×10^4 J

    ANS: D   DIF: IIIC   OBJ: 6-3.2

52. A clay ball with a mass of 0.35 kg has an initial speed of 4.2 m/s. It strikes a 3.5 kg clay ball at rest, and the two balls stick together and remain stationary. What is the decrease in kinetic energy of the 0.35 kg ball?
    a. 1.6 J
    c. 3.1 J
    b. 4.8 J
    d. 6.4 J

    ANS: C   DIF: IIIC   OBJ: 6-3.2

53. An infant throws 5 g of applesauce at a velocity of 0.2 m/s. All of the applesauce collides with a nearby wall and sticks. What is the decrease in kinetic energy of the applesauce?
    a. 2×10^{-4} J
    c. 1×10^{-3} J
    b. 0.5×10^{-4} J
    d. 1×10^{-4} J

    ANS: D   DIF: II   OBJ: 6-3.2

54. In an elastic collision between two objects with unequal masses,
    a. the total momentum of the system will increase.
    b. the total momentum of the system will decrease.
    c. the kinetic energy of one object will increase by the amount that the kinetic energy of the other object decreases.
    d. the momentum of one object will increase by the amount that the momentum of the other object decreases.

    ANS: D   DIF: IIIB   OBJ: 6-3.3

55. A billiard ball collides with a stationary identical billiard ball in an elastic head-on collision. After the collision, which is true of the first ball?
    a. It maintains its initial velocity.   c. It comes to rest.
    b. It has one-half its initial velocity.   d. It moves in the opposite direction.

    ANS: C   DIF: I   OBJ: 6-3.3

56. A billiard ball collides with a second identical ball in an elastic head-on collision. What is the kinetic energy of the system after the collision compared with the kinetic energy before the collision?
    a. unchanged   c. two times as great
    b. one-fourth as great   d. four times as great

    ANS: A   DIF: I   OBJ: 6-3.3

57. Which of the following best describes the kinetic energy of each object after a two-body collision if the momentum of the system is conserved?
    a. must be less   c. might also be conserved
    b. must also be conserved   d. is doubled in value

    ANS: C   DIF: I   OBJ: 6-3.3

58. Which of the following best describes the momenta of two bodies after a two-body collision if the kinetic energy of the system is conserved?
    a. must be less   c. might also be conserved
    b. must also be conserved   d. is doubled in value

    ANS: B   DIF: I   OBJ: 6-3.3

59. An object with a mass of 0.10 kg makes an elastic head-on collision with a stationary object with a mass of 0.15 kg. The final velocity of the 0.10 kg object after the collision is –0.045 m/s. What was the initial velocity of the 0.10 kg object?
    a. 0.16 m/s   c. 0.20 m/s
    b. –1.06 m/s   d. –0.20 m/s

    ANS: C   DIF: IIIC   OBJ: 6-3.4

60. A bowling ball with a mass of 7.0 kg strikes a pin that has a mass of 2.0 kg. The pin flies forward with a velocity of 6.0 m/s, and the ball continues forward at 4.0 m/s. What was the original velocity of the ball?
    a. 4.0 m/s
    b. 5.7 m/s
    c. 6.6 m/s
    d. 3.3 m/s

    ANS: B   DIF: IIIB   OBJ: 6-3.4

61. A 90 kg halfback runs north and is tackled by a 120 kg opponent running south at 4 m/s. The collision is perfectly inelastic. Just after the tackle, both players move at a velocity of 2 m/s north. Calculate the velocity of the 90 kg player just before the tackle.
    a. 3 m/s south
    b. 4 m/s south
    c. 10 m/s north
    d. 12 m/s north

    ANS: C   DIF: IIIB   OBJ: 6-3.4

62. A clay ball with a mass of 0.35 kg strikes another 0.35 kg clay ball at rest, and the two balls stick together. The final velocity of the balls is 2.1 m/s north. What was the first ball's initial velocity?
    a. 4.2 m/s to the north
    b. 2.1 m/s to the south
    c. 2.1 m/s to the north
    d. 4.2 m/s to the south

    ANS: A   DIF: IIIB   OBJ: 6-3.4

63. A 2 kg mass moving to the right makes an elastic head-on collision with a 4 kg mass moving to the left at 4 m/s. The 2 kg mass reverses direction after the collision and moves at 3 m/s. The 4 kg mass moves to the left at 1 m/s. What was the initial velocity of the 4 kg mass?
    a. 3 m/s to the right
    b. 1 m/s to the left
    c. 4 m/s to the left
    d. 4 m/s to the right

    ANS: A   DIF: IIIB   OBJ: 6-3.4

64. A 15 g marble moves to the right at 3.5 m/s and makes an elastic head-on collision with a 22 g marble. The final velocity of the 15 g marble is 5.4 m/s to the left, and the final velocity of the 22 g marble is 2.0 m/s to the right. What is the initial velocity of the 22 g marble?
    a. 5.3 m/s to the left
    b. 5.3 m/s to the right
    c. 4.0 m/s to the left
    d. 4.0 m/s to the right

    ANS: C   DIF: IIIC   OBJ: 6-3.4

SHORT ANSWER

1. On a pool table, a moving cue ball collides with the eight ball, which is at rest. Is it possible for both balls to be at rest after the collision? Use the law of conservation of momentum to explain your reasoning.

   ANS:
   No, the final momentum can equal zero only if the initial momentum was zero. Because the cue ball was moving, its initial momentum was not zero. Therefore, both balls cannot be at rest after the collision.

   DIF: II   OBJ: 6-2.3

MULTIPLE CHOICE

1. Which of the following angles equals 2π rad?
   a. 360°
   b. 180°
   c. 0°
   d. 3.14°

   ANS: A　　　　DIF: I　　　　OBJ: 7-1.1

2. One radian is equal to
   a. 60°.
   b. 58°.
   c. 57.3°.
   d. 56°.

   ANS: C　　　　DIF: IIIA　　　　OBJ: 7-1.1

3. How would an angle in radians be converted to an angle in degrees?
   a. The angle in radians would be multiplied by $180°/\pi$.
   b. The angle in radians would be multiplied by $360°/\pi$.
   c. The angle in radians would be multiplied by $180°/2\pi$.
   d. The angle in radians would be multiplied by $2\pi/360°$.

   ANS: A　　　　DIF: II　　　　OBJ: 7-1.1

4. How would you convert an angle in degrees to an angle in radians?
   a. multiply the angle measured in degrees by $2\pi/180°$
   b. multiply the angle measured in degrees by $2\pi/360°$
   c. multiply the angle measured in degrees by $\pi/360°$
   d. multiply the angle measured in degrees by $2\pi r°$

   ANS: B　　　　DIF: I　　　　OBJ: 7-1.1

5. A cave dweller rotates a pebble in a sling with a radius of 0.30 m counterclockwise through an arc length of 0.96 m. What is the angular displacement of the pebble?
   a. 1.6 rad
   b. −1.6 rad
   c. 3.2 rad
   d. −3.2 rad

   ANS: C　　　　DIF: IIIB　　　　OBJ: 7-1.2

6. Earth has an equatorial radius of approximately 6380 km, and it rotates 360° every 24 h. What is the angular displacement of a person standing at the equator for 3.0 h?
   a. 0.26 rad
   b. 0.52 rad
   c. 0.78 rad
   d. 0.39 rad

   ANS: C　　　　DIF: IIIA　　　　OBJ: 7-1.2

7. A child sits on a carousel at a distance of 3.5 m from the center and rotates through an arc length of 6.5 m. What is the angular displacement of the child?
   a. 1.9 rad
   b. 0.93 rad
   c. 3.0 rad
   d. 5.0 rad

   ANS: A   DIF: IIIB   OBJ: 7-1.2

8. A bucket on the circumference of a water wheel travels an arc length of 18 m. If the radius of the wheel is 4.1 m, what is the angular displacement of the bucket?
   a. 1.0 rad
   b. 4.4 rad
   c. 3.7 rad
   d. 2.3 rad

   ANS: B   DIF: IIIB   OBJ: 7-1.2

9. What is the approximate angular speed of a wheel rotating at the rate of 5.0 rev/s?
   a. 3.2 rad/s
   b. 1.6 rad/s
   c. 16 rad/s
   d. 31 rad/s

   ANS: D   DIF: IIIA   OBJ: 7-1.3

10. A grinding wheel initially at rest with a radius of 0.15 m rotates until it reaches an angular speed of 12.0 rad/s in 4.0 s. What is the wheel's average angular acceleration?
    a. 96 rad/s^2
    b. 48 rad/s^2
    c. 3.0 rad/s^2
    d. 0.33 rad/s^2

    ANS: C   DIF: IIIA   OBJ: 7-1.3

11. A potter's wheel moves from rest to an angular speed of 0.54 rad/s in 30.0 s. What is the angular acceleration of the wheel?
    a. 16 rad/s^2
    b. 1.3 rad/s^2
    c. 0.018 rad/s^2
    d. 0.042 rad/s^2

    ANS: C   DIF: IIIB   OBJ: 7-1.3

12. A record player is turned on and reaches an angular velocity of 4.7 rad/s in 1.37 s. What is the average angular acceleration of the record?
    a. 3.4 rad/s^2
    b. 4.3 rad/s^2
    c. 6.4 rad/s^2
    d. 0.29 rad/s^2

    ANS: A   DIF: IIIA   OBJ: 7-1.3

13. A Ferris wheel initially at rest accelerates to a final angular speed of 0.70 rad/s and rotates through an angular displacement of 4.90 rad. What is the Ferris wheel's average angular acceleration?
    a. 0.10 rad/s^2
    b. 0.05 rad/s^2
    c. 1.80 rad/s^2
    d. 0.60 rad/s^2

    ANS: B   DIF: IIIB   OBJ: 7-1.3

14. A Ferris wheel rotates with an initial angular speed of 0.50 rad/s and accelerates over a 7.00 s interval at a rate of 4.0×10^{-2} rad/s². What is its angular speed?
    a. 0.20 rad/s
    b. 0.30 rad/s
    c. 0.46 rad/s
    d. 0.78 rad/s

    ANS: D   DIF: IIIB   OBJ: 7-1.4

15. An automobile tire with a radius of 0.30 m starts at rest and accelerates at a constant angular acceleration of 2.0 rad/s² for 5.0 s. What is the angular displacement of the tire?
    a. 12 rad
    b. 25 rad
    c. 2.0 rad
    d. 0.50 rad

    ANS: B   DIF: I   OBJ: 7-1.4

16. A bicycle wheel rotates with a constant angular acceleration of 3.0 rad/s². If the initial angular speed of the wheel is 1.5 rad/s, what is the angular displacement of the wheel after 4.0 s?
    a. 6.0 rad
    b. 24 rad
    c. 3.0×10^1 rad
    d. 36 rad

    ANS: C   DIF: IIIB   OBJ: 7-1.4

17. A gear in a machine accelerates at 11.2 rad/s². If the wheel's initial angular speed is 5.40 rad/s, what is the wheel's angular speed after exactly 3.0 seconds?
    a. 39.0 rad/s
    b. 13.6 rad/s
    c. 209 rad/s
    d. 28.2 rad/s

    ANS: A   DIF: IIIB   OBJ: 7-1.4

18. A ball rolls downhill with an angular speed of 2.5 rad/s and has a constant angular acceleration of 2.0 rad/s². If the ball takes 11.5 s to reach the bottom of the hill, what is the final angular speed of the ball?
    a. 13 rad
    b. 31 rad/s
    c. 33 rad/s
    d. 25.5 rad/s

    ANS: D   DIF: IIIB   OBJ: 7-1.4

19. A helicopter has 3.0 m long rotor blades that are rotating at an angular speed of 63 rad/s. What is the tangential speed of each blade tip?
    a. 99 m/s
    b. 190 m/s
    c. 21 m/s
    d. 66 m/s

    ANS: B   DIF: IIIC   OBJ: 7-2.1

20. The end of the cord on a weed cutter is 0.15 m in length. If the motor rotates at the rate of 126 rad/s, what is the tangential speed of the cord?
    a. 628 m/s
    b. 25 m/s
    c. 19 m/s
    d. 63 m/s

    ANS: C   DIF: IIIB   OBJ: 7-2.1

21. A point on the rim of a 0.30 m radius rotating wheel has a tangential speed of 4.0 m/s. What is the tangential speed of a point 0.20 m from the center of the same wheel?
   a. 0.8 m/s
   b. 1.3 m/s
   c. 2.6 m/s
   d. 8.0 m/s

   ANS: C   DIF: IIIA   OBJ: 7-2.1

22. A cylinder with a diameter of 0.150 m rotates in a lathe at a constant angular speed of 35.6 rad/s. What is the tangential speed of the surface of the cylinder?
   a. 2.67 m/s
   b. 5.34 m/s
   c. 2.37×10^2 m/s
   d. 4.75×10^2 m/s

   ANS: B   DIF: IIIA   OBJ: 7-2.1

23. A wheel with a radius of 1.2 m rotates at a constant angular speed of 10.5 rad/s. What is the tangential speed of a point 0.55 m from the wheel's axis?
   a. 19 m/s
   b. 5.8 m/s
   c. 13 m/s
   d. 8.7 m/s

   ANS: B   DIF: IIIC   OBJ: 7-2.1

24. An automobile tire with a radius of 0.3 m accelerates from rest at a constant 2 rad/s^2 over a 5 s interval. What is the tangential component of acceleration for a point on the outer edge of the tire?
   a. 30 m/s^2
   b. 7 m/s^2
   c. 0.6 m/s^2
   d. 0.3 m/s^2

   ANS: C   DIF: IIIB   OBJ: 7-2.2

25. A hamster gets on a stationary wheel with a radius of 0.15 m and runs until the wheel rotates at an angular speed of 12.0 rad/s in 4.0 s. What is the tangential acceleration of the wheel's edge?
   a. 0.45 rad/s^2
   b. 0.6 rad/s^2
   c. 0.65 rad/s^2
   d. 1.30 rad/s^2

   ANS: A   DIF: IIIB   OBJ: 7-2.2

26. A flywheel with a radius of 0.30 m starts from rest and accelerates with a constant angular acceleration of 0.50 rad/s^2. What is the tangential acceleration of the flywheel?
   a. 0.63 m/s^2
   b. 0.15 m/s^2
   c. 0.65 m/s^2
   d. 1.30 m/s^2

   ANS: B   DIF: IIIC   OBJ: 7-2.2

27. A contestant in a game show spins a stationary wheel with a radius of 0.50 m so that it has a constant angular acceleration of 0.40 rad/s^2. What is the tangential acceleration of a point on the edge of the wheel?
   a. 0.20 m/s^2
   b. 0.60 m/s^2
   c. 1.3 m/s^2
   d. 0.73 m/s^2

   ANS: A   DIF: IIIB   OBJ: 7-2.2

28. A stone on the edge of the tire of a unicycle wheel with a radius of 0.25 m has a centripetal acceleration of 4.0 m/s². What is the tire's angular speed?
   a.   1.0 rad/s
   b.   2.0 rad/s
   c.   3.2 rad/s
   d.   4.0 rad/s

   ANS: D   DIF: IIIB   OBJ: 7-2.3

29. A point on the rim of a rotating wheel with a 0.37 m radius has a centripetal acceleration of 19.0 m/s². What is the angular speed of the wheel?
   a.   0.89 m/s
   b.   1.6 rad/s
   c.   3.2 rad/s
   d.   7.2 rad/s

   ANS: D   DIF: IIIB   OBJ: 7-2.3

30. A lapidary plate at rest is turned on to cut a gemstone. The plate rotates until it reaches an angular speed of 12.0 rad/s in 4.0 s. What is the centripetal acceleration of a point 0.10 m from the center of the plate?
   a.   0.45 m/s²
   b.   7.2 m/s²
   c.   14 m/s²
   d.   29 m/s²

   ANS: C   DIF: IIIB   OBJ: 7-2.3

31. If the distance from the center of a merry-go-round to the edge is 1.2 m, what centripetal acceleration does a passenger experience when the merry-go-round rotates at an angular speed of 0.5 rad/s?
   a.   1.7 m/s²
   b.   0.9 m/s²
   c.   0.3 m/s²
   d.   0.6 m/s²

   ANS: C   DIF: IIIB   OBJ: 7-2.3

32. A 80.0 kg passenger is seated 12 m from the center of the loop of a roller coaster. What centripetal force does the passenger experience when the roller coaster reaches an angular speed of 3.14 rad/s?
   a. 1.7×10^3 N
   b. 6.9×10^3 N
   c. 7.2×10^3 N
   d. 9.5×10^3 N

   ANS: D   DIF: IIIA   OBJ: 7-3.1

33. A 0.40 kg ball on a 0.50 m string rotates in a circular path in a vertical plane. If the angular speed of the ball at the bottom of the circle is 8.0 rad/s, what is the force that maintains circular motion?
   a.   5.6 N
   b.   11 N
   c.   13 N
   d.   20.0 N

   ANS: C   DIF: IIIC   OBJ: 7-3.1

34. A 0.40 kg ball on a 0.50 m string rotates in a circular path in a vertical plane. If a constant angular speed of 8.0 rad/s is maintained, what is the tension in the string when the ball is at the top of the circle?
   a.   9.0 N
   b.   11 N
   c.   13 N
   d.   10.0 N

   ANS: A   DIF: IIIC   OBJ: 7-3.1

35. A roller coaster loaded with passengers has a mass of 2.0×10^3 kg; the radius of curvature of the track at the lowest point of the track is 24 m. If the vehicle has a tangential speed of 18 m/s at this point, what force is exerted on the vehicle by the track?
   a. 2.3×10^4 N
   b. 4.7×10^4 N
   c. 3.0×10^4 N
   d. 2.7×10^4 N

   ANS: D   DIF: IIIB   OBJ: 7-3.1

36. What is the gravitational force between two trucks, each with a mass of 2.0×10^4 kg, that are 2.0 m apart? ($G = 6.673 \times 10^{-11}$ N•m^2/kg^2)
   a. 5.7×10^{-2} N
   b. 1.3×10^{-2} N
   c. 6.7×10^{-3} N
   d. 1.2×10^{-7} N

   ANS: C   DIF: IIIB   OBJ: 7-3.3

37. The gravitational force between two masses is 36 N. What is the gravitational force if the distance between them is tripled? ($G = 6.673 \times 10^{-11}$ N•m^2/kg^2)
   a. 4.0 N
   b. 9.0 N
   c. 18 N
   d. 27 N

   ANS: A   DIF: IIIA   OBJ: 7-3.3

38. Two small masses that are 10.0 cm apart attract each other with a force of 10.0 N. When they are 5.0 cm apart, these masses will attract each other with what force?
   ($G = 6.673 \times 10^{-11}$ N•m^2/kg^2)
   a. 5.0 N
   b. 2.5 N
   c. 20.0 N
   d. 40.0 N

   ANS: D   DIF: II   OBJ: 7-3.3

SHORT ANSWER

1. Is there an outward force in circular motion?

   ANS:
   No, there is only an inward force causing a deviation from a straight-line path.

   DIF: I   OBJ: 7-3.2

2. Pizza makers traditionally form the crust by throwing the dough in the air and spinning it. Why does this make the pizza crust bigger?

   ANS:
   Each point on the crust (except the center) has an inertial tendency to move in a straight line. This causes the circle to become bigger as the points move away from each other. The cohesion of the dough keeps it from flying apart.

   DIF: I   OBJ: 7-3.2

3. A ball is whirled in a horizontal circular path on the end of a string. Predict the path of the ball when the string breaks, and explain your answer.

ANS:
Inertia causes the ball to move in a straight path tangent to the circle.

DIF: I   OBJ: 7-3.2

4. A parent holds a child by the arms and spins around in a circle at a constant speed. If the parent spins fast enough, will the child's feet leave the ground? Explain your answer.

ANS:
As the angular velocity increases, the parent's arms must exert a larger and larger force, F, because the horizontal component of this force, F_h, is the centripetal force and this force $F_c = mr\omega^2$. However, if F increases so does its vertical component. When F is large enough so that its vertical component is equal to the weight of the child, the child's feet leave the ground.

DIF: I   OBJ: 7-3.2

5. A tether ball is tied to a string and whirled in a horizontal circular path at a constant speed. What causes the ball and string to move away from the post?

ANS:
The tension in the string is the force that maintains the circular motion and acts along the string away from the ball. Inertia causes the ball and string to move outward.

DIF: I   OBJ: 7-3.2

PROBLEM

1. A 61 kg student sits at a desk 1.25 m away from a 70.0 kg student. What is the magnitude of the gravitational force between the two students?
($G = 6.673 \times 10^{-11}$ N•m²/kg²)

ANS:
1.8×10^{-7} N

DIF: IIIB   OBJ: 7-3.3

MULTIPLE CHOICE

1. If a net torque is applied to an object, that object will experience which of the following?
   a. a constant angular speed
   b. an angular acceleration
   c. a constant moment of inertia
   d. an increasing moment of inertia

   ANS: B   DIF: I   OBJ: 8-1.2

2. Which of the following quantities measures the ability of a force to rotate or accelerate an object around an axis?
   a. axis of rotation
   b. lever arm
   c. moment arm
   d. torque

   ANS: D   DIF: I   OBJ: 8-1.2

3. Which of the following statements is correct?
   a. The farther the force is from the axis of rotation, the more torque is produced.
   b. The closer the force is to the axis of rotation, the more torque is produced.
   c. The closer the force is to the axis of rotation, the easier it is to rotate the object.
   d. The farther the force is from the axis of rotation, the less torque is produced.

   ANS: A   DIF: I   OBJ: 8-1.2

4. Where should a force be applied on a lever arm to produce the most torque?
   a. closest to the axis of rotation
   b. farthest from the axis of rotation
   c. in the middle of the lever arm
   d. It doesn't matter where the force is applied.

   ANS: B   DIF: I   OBJ: 8-1.2

5. To warm up before a game, a baseball pitcher tosses a 0.15 kg ball by rotating his forearm, which is 0.32 m in length, to accelerate the ball. The ball starts at rest and is thrown at a speed of 12 m/s in 0.40 s. While the ball is in the pitcher's hand, what torque is applied to the ball to produce the angular acceleration?
   a. 1.1 N•m
   b. 11 N•m
   c. 7.2 N•m
   d. 1.4 N•m

   ANS: D   DIF: IIIB   OBJ: 8-1.3

6. A bucket filled with water has a mass of 23 kg and is attached to a rope that is wound around a cylinder with a radius of 0.050 m at the top of a well. What torque does the weight of the water and bucket produce on the cylinder? ($g = 9.81$ m/s^2.)
   a. 34 N•m
   b. 17 N•m
   c. 11 N•m
   d. 23 N•m

   ANS: C   DIF: IIIB   OBJ: 8-1.3

7. A force of 4.0 N is applied to a door at an angle of 60.0° and a distance of 0.30 m from the hinge. What is the torque produced?
   a. 1.0 N•m
   b. 0.75 N•m
   c. 0.87 N•m
   d. 0.22 N•m

   ANS: A   DIF: IIIB   OBJ: 8-1.3

8. A heavy bank-vault door is opened by the application of a force of 3.0×10^2 N directed perpendicular to the plane of the door at a distance of 0.80 m from the hinges. What is the torque?
   a. 120 N•m
   b. 240 N•m
   c. 300 N•m
   d. 360 N•m

   ANS: B   DIF: IIIB   OBJ: 8-1.3

9. Suppose a doorknob is placed at the center of a door. Compared with a door whose knob is located at the edge, what amount of force must be applied to this door to produce the torque exerted on the other door?
   a. one-half as much
   b. two times as much
   c. one-fourth as much
   d. four times as much

   ANS: B   DIF: II   OBJ: 8-1.4

10. If you want to open a swinging door with the least amount of force, where should you push on the door?
    a. close to the hinges
    b. in the middle
    c. as far from the hinges as possible
    d. It does not matter where you push.

    ANS: C   DIF: I   OBJ: 8-1.4

11. If you cannot exert enough force to loosen a bolt with a wrench, which of the following should you do?
    a. Use a wrench with a longer handle.
    b. Tie a rope to the end of the wrench and pull on the rope.
    c. Use a wrench with a shorter handle.
    d. You should exert a force on the wrench closer to the bolt.

    ANS: A   DIF: II   OBJ: 8-1.4

12. If the torque required to loosen a nut on a wheel has a magnitude of 40.0 N•m and the force exerted by a mechanic is 133 N, how far from the nut must the mechanic apply the force?
    a. 60.0 cm
    b. 15.0 cm
    c. 30.0 cm
    d. 1.20 m

    ANS: C   DIF: IIIA   OBJ: 8-1.4

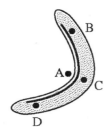

13. At which point in the figure above is the approximate center of mass?
   a. A
   b. B
   c. C
   d. D

   ANS: A   DIF: II   OBJ: 8-2.1

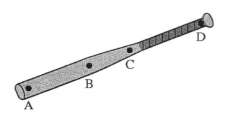

14. At which point on the baseball bat above is the approximate center of mass?
   a. A
   b. B
   c. C
   d. D

   ANS: B   DIF: II   OBJ: 8-2.1

15. At which point on the hammer above is the approximate center of mass?
   a. A
   b. B
   c. C
   d. D

   ANS: D   DIF: II   OBJ: 8-2.1

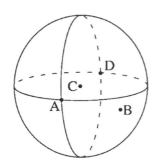

16. At which point on the hollow sphere above is the approximate center of mass?
    a. A
    b. B
    c. C
    d. D

    ANS: C   DIF: II   OBJ: 8-2.1

17. Which of the following is NOT an intrinsic property of an object?
    a. mass
    b. moment of inertia
    c. center of mass
    d. center of gravity

    ANS: B   DIF: I   OBJ: 8-2.2

18. Which of the following statements is correct?
    a. The farther the center of mass of an object is from the axis of rotation, the less difficult it is to rotate the object.
    b. The farther the center of mass of an object is from the axis of rotation, the smaller the object's moment of inertia is.
    c. The farther the center of mass of an object is from the axis of rotation, the greater the object's moment of inertia is.
    d. The farther the center of mass of an object is from the axis of rotation, the greater the object's moment of inertia is, but the less difficult it is to rotate the object.

    ANS: C   DIF: I   OBJ: 8-2.2

19. The dependence of equilibrium on the absence of net torque is
    a. the first condition of equilibrium.
    b. the second condition of equilibrium.
    c. rotational equilibrium.
    d. translational equilibrium.

    ANS: B   DIF: I   OBJ: 8-2.3

20. A uniform bridge span weighs 5.00×10^4 N and is 40.0 m long. An automobile weighing 1.50×10^4 N is parked with its center of gravity located 12.0 m from the right pier. What upward support force is provided by the left pier?
    a. 2.95×10^4 N
    b. 3.55×10^4 N
    c. 6.50×10^4 N
    d. 3.25×10^4 N

    ANS: A   DIF: IIIB   OBJ: 8-2.4

21. A meterstick supported by a knife edge at the 50 cm mark has masses of 0.40 kg and 0.60 kg hanging from the 20 cm and 80 cm marks, respectively. At what mark should a third mass of 0.30 kg be hung to keep the stick balanced?
   a.   20 cm
   b.   70 cm
   c.   30 cm
   d.   25 cm

   ANS: C   DIF: IIIB   OBJ: 8-2.4

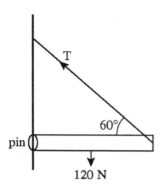

22. A uniform horizontal beam with a length of 6.0 m and a weight of 120 N is attached at one end to a wall by a pin connection so that the beam can rotate. The opposite end of the beam is supported by a cable attached to the wall above the pin. The cable makes an angle of 60.0° with the beam. What is the tension in the cable needed to maintain the beam in equilibrium?
   a.   35 N
   b.   69 N
   c. 6.0×10^1 N
   d.   120 N

   ANS: B   DIF: IIIB   OBJ: 8-2.4

23. A child with a weight of 4.50×10^2 N sits on a seesaw 0.60 m from the axis of rotation. How far from the axis of rotation on the other side should a child with a weight of 6.00×10^2 N sit so the seesaw will remain balanced?
   a.   0.30 m
   b.   0.40 m
   c.   0.45 m
   d.   0.50 m

   ANS: C   DIF: IIIA   OBJ: 8-2.4

24. According to Newton's second law, the angular acceleration experienced by an object is directly proportional to which of the following?
   a.   the object's moment of inertia
   b.   the net applied torque
   c.   the size of the object
   d.   the mass of the object

   ANS: B   DIF: I   OBJ: 8-3.1

25. Which of the following statements is correct?
   a.   With a net positive torque, the angular acceleration of an object is clockwise.
   b.   With a net positive torque, the angular acceleration of an object is counterclockwise.
   c.   With a net negative torque, the angular acceleration of an object is counterclockwise.
   d.   The net force of an object is not related to the translational acceleration given to the object.

   ANS: B   DIF: I   OBJ: 8-3.1

26. Which of the following represents Newton's second law for rotating objects?
    a. net torque = moment of inertia × angular acceleration
    b. net torque = moment of inertia ÷ angular acceleration
    c. force = mass × acceleration
    d. force = mass ÷ acceleration

    ANS: A   DIF: I   OBJ: 8-3.1

27. A grinding wheel with a moment of inertia of 2.0 kg•m^2 is initially at rest. What angular momentum will the wheel have 10.0 s after a 2.5 N•m torque is applied to it?
    a. 25 kg•m^2/s
    b. 7.5 kg•m^2/s
    c. 4.0 kg•m^2/s
    d. 0.25 kg•m^2/s

    ANS: A   DIF: I   OBJ: 8-3.2

28. A bowling ball has a mass of 7.0 kg, a moment of inertia of 2.8×10^{-2} kg•m^2, and a radius of 0.10 m. If it rolls down the lane without slipping at an angular speed of 4.0×10^1 rad/s, what is its angular momentum?
    a. 0.80 kg•m^2/s
    b. 1.4 kg•m^2/s
    c. 11 kg•m^2/s
    d. 1.1 kg•m^2/s

    ANS: D   DIF: IIIA   OBJ: 8-3.2

29. A figure skater with arms drawn in spins on the ice at a rate of 5.0 rad/s and has a moment of inertia of 1.875 kg•m^2. What is the angular momentum of the skater?
    a. 2.5 kg•m^2/s
    b. 3.8 kg•m^2/s
    c. 9.4 kg•m^2/s
    d. 12 kg•m^2/s

    ANS: C   DIF: IIIA   OBJ: 8-3.2

30. The moment of inertia of a cylinder is 0.016 kg•m^2. If the angular speed is 15.7 rad/s, what is the angular momentum of the cylinder?
    a. 0.25 kg•m^2/s
    b. 12.1 kg•m^2/s
    c. 19.2 kg•m^2/s
    d. 28.6 kg•m^2/s

    ANS: A   DIF: IIIB   OBJ: 8-3.2

31. A 2.50 N•m torque is applied to a grinding wheel that has a moment of inertia of 2.00 kg•m^2. What is the final kinetic energy of the grinding wheel 10.0 s after beginning from rest?
    a. 312 J
    b. 237 J
    c. 156 J
    d. 106 J

    ANS: C   DIF: IIIB   OBJ: 8-3.3

32. The total kinetic energy of a baseball thrown with a spinning motion is a function of which of the following?
   a. the ball's linear speed only
   b. the ball's rotational speed only
   c. both the ball's linear and rotational speeds
   d. neither the ball's linear nor the ball's rotational speed

   ANS: C   DIF: I   OBJ: 8-3.3

33. A bowling ball has a mass of 7.0 kg, a moment of inertia of 2.8×10^{-2} kg•m^2, and a radius of 0.10 m. If it rolls down the lane without slipping at a linear speed of 4.0 m/s, what is its total kinetic energy?
   a. 45 J   c. 11 J
   b. 32 J   d. 78 J

   ANS: D   DIF: IIIB   OBJ: 8-3.3

34. A bucket filled with water has a mass of 23 kg and is attached to a rope that is wound with a crank around a 0.05 m radius cylinder at the top of a well. The moment of inertia of the cylinder and crank is 0.12 kg•m^2. The bucket and water are first raised to the top of the well and then released to fall back into the well. What is the rotational kinetic energy of the cylinder and crank at the instant the bucket is moving at a speed of 7.9 m/s?
   a. 2.1×10^3 J   c. 7.0×10^2 J
   b. 1.5×10^3 J   d. 4.0×10^2 J

   ANS: B   DIF: IIIC   OBJ: 8-3.3

35. A solid sphere with a mass of 4.0 kg and a radius of 0.12 m starts from rest at the top of a ramp inclined at 15° and rolls to the bottom. The upper end of the ramp is 1.2 m higher than the lower end. What is the total kinetic energy of the sphere when it reaches the bottom? (Assume that the sphere rolls without slipping and that $g = 9.81$ m/s^2.)
   a. 70 J   c. 18 J
   b. 47 J   d. 8.8 J

   ANS: B   DIF: IIIA   OBJ: 8-3.3

SHORT ANSWER

1. Should a bowling ball rolling toward pins be treated as a point mass or an extended object?

   ANS:
   extended object

   DIF: I   OBJ: 8-1.1

2. Should a pin being hit by a bowling ball be treated as a point mass or an extended object?

   ANS:
   extended object

   DIF: I   OBJ: 8-1.1

3. Should a tennis ball rolling toward the net be treated as a point mass or an extended object?

ANS:
extended object

DIF: I   OBJ: 8-1.1

4. Should a tennis ball lobbed over the net be treated as a point mass or an extended object?

ANS:
point mass

DIF: I   OBJ: 8-1.1

5. What occurs when a force is applied to the edge of an extended object?

ANS:
The force produces a torque, which causes the object to rotate.

DIF: I

6. Where is the center of mass of a meterstick?

ANS:
at the 50.0 cm mark

DIF: I   OBJ: 8-2.1

7. What is the main difference between mass and moment of inertia?

ANS:
Mass resists changes in translational motion, and moment of inertia resists changes in rotational motion.

DIF: I   OBJ: 8-2.2

8. Define the second condition of equilibrium.

ANS:
The second condition of equilibrium is the dependence of equilibrium on the absence of net torque.

DIF: I   OBJ: 8-2.3

9. How can an object be in both rotational and translational equilibrium?

ANS:
There must be zero net force and zero net torque.

DIF: I   OBJ: 8-2.3

10. Describe Newton's second law of rotation.

   ANS:
   Net torque is equal to an object's moment of inertia multiplied by its angular acceleration.

   OBJ: 8-3.1

11. Identify the simple machine in the figure above.

   ANS:
   inclined plane

   DIF: I   OBJ: 8-4.1

12. Identify the simple machine in the figure above.

   ANS:
   wedge

   DIF: I   OBJ: 8-4.1

13. Identify the simple machine in the figure above.

   ANS:
   screw

   DIF: I   OBJ: 8-4.1

14. Identify the simple machine in the figure above.

    ANS:
    lever

    DIF: I   OBJ: 8-4.1

15. Identify the simple machine in the figure above.

    ANS:
    wheel and axle (or pulley)

    DIF: I   OBJ: 8-4.1

16. Explain how the operation of a simple machine alters the applied force and the distance moved.

    ANS:
    A machine can increase or decrease the force acting on an object at the expense or gain of the distance moved.

    DIF: I   OBJ: 8-4.2

17. How does the use of a machine alter the work done on the object?

    ANS:
    The work done on the object is constant.

    DIF: I   OBJ: 8-4.2

18. When a machine increases the force acting on an object, what happens to the distance the object moved?

    ANS:
    The distance moved decreases.

    DIF: I   OBJ: 8-4.2

19. When a machine decreases the force acting on an object, what happens to the distance the object moved?

ANS:
The distance moved increases.

DIF: I   OBJ: 8-4.2

PROBLEM

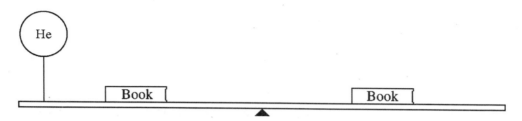

1. A 4.0 m board with a mass of 19 kg is pivoted at its center of gravity. A helium balloon attached 0.23 m from the left end of the board produces an upward force of 7.0 N. A 3.5 kg book is placed 0.73 m from the left end of the board, and another book of 1.3 kg is placed 0.75 m from the right end of the board. Find the torque on the board and the direction of rotation.

ANS:
15 N•m counterclockwise

DIF: IIIB   OBJ: 8-1.3

2. A girl pushes a box that has a mass of 650 N up an incline. If the girl exerts a force of 150 N along the incline, what is the mechanical advantage of the incline?

ANS:
4.3

DIF: IIIB   OBJ: 8-4.3

3. An iron bar is used to lift a slab of cement. The force applied to lift the slab is 4.0×10^2 N. If the slab weighs 6400 N, what is the mechanical advantage of the bar?

ANS:
16

DIF: IIIB   OBJ: 8-4.3

4. A force of 255 N is needed to pull a nail from a wall using a claw hammer. If the resistance force of the nail is 3650 N, what is the mechanical advantage of the hammer?

ANS:
14.3

DIF: IIIB   OBJ: 8-4.3

5. A boy can raise a rock that weighs 95 N by using a lever and applying a force of 15 N. What is the mechanical advantage of the lever?

ANS:
6.3

DIF: IIIB   OBJ: 8-4.3

6. What is the efficiency of a machine that requires 1.00×10^2 J of input energy to do 35 J of work?

ANS:
35%

DIF: IIIA   OBJ: 8-4.4

7. A box weighing 210 N is pushed up an inclined plane that is 2.0 m long. A force of 140 N is required. If the box is lifted 1.0 m, what is the efficiency of the inclined plane?

ANS:
75%

DIF: IIIB   OBJ: 8-4.4

8. A force of 1250 N is needed to move a crate weighing 3270 N up a ramp that is 4.55 m long. If the elevated end of the ramp is 0.750 m high, what is the efficiency of the ramp?

ANS:
43.1%

DIF: IIIB   OBJ: 8-4.4

9. What is the efficiency of a machine that requires 135 J of input energy to do 87.5 J of work?

ANS:
64.8%

DIF: IIIA   OBJ: 8-4.4

10. What is the efficiency of a machine that requires 1463 J of input energy to do 670 J of work?

ANS:
46%

DIF: IIIA   OBJ: 8-4.4

MULTIPLE CHOICE

1. Which of the following is a fluid?
   a. helium
   b. ice
   c. iron
   d. gold

   ANS: A   DIF: I   OBJ: 9-1.1

2. Which of the following is NOT a fluid?
   a. carbon dioxide
   b. hydrogen
   c. sea water
   d. glass

   ANS: D   DIF: I   OBJ: 9-1.1

3. Which of the following statements is NOT correct?
   a. A fluid flows.
   b. A fluid has a definite shape.
   c. Molecules of a fluid are free to move past each other.
   d. A fluid changes its shape easily.

   ANS: B   DIF: I   OBJ: 9-1.1

4. A table-tennis ball has an average density of 0.084 g/cm^3 and a diameter of 3.8 cm. What force can submerge the ball in water? ($\rho_w = 1.00$ g/cm^3)
   a. 1.0 N
   b. 0.79 N
   c. 0.52 N
   d. 0.26 N

   ANS: D   DIF: IIIA   OBJ: 9-1.3

5. A cube of wood with a density of 0.780 g/cm^3 is 10.0 cm on each side. When the cube is placed in water, what buoyant force acts on the wood? ($\rho_w = 1.00$ g/cm^3)
   a. 7.65×10^3 N
   b. 5.00 N
   c. 7.65 N
   d. 6.40 N

   ANS: C   DIF: IIIA   OBJ: 9-1.3

6. According to legend, to determine whether the king's crown was made of pure gold, Archimedes measured the crown's volume by determining how much water it displaced. The density of gold is 19.3 g/cm^3. If the crown's mass was 6.00×10^2 g, what volume of water would have been displaced if the crown was indeed made of pure gold?
   a. 31.1 cm^3
   b. 114×10^3 cm^3
   c. 22.8×10^3 cm^3
   d. 1.81×10^3 cm^3

   ANS: A   DIF: IIIB   OBJ: 9-1.3

7. Because a buoyant force acts in the opposite direction of gravity,
   a. objects submerged in water have a net force smaller than their weight.
   b. objects submerged in water have a net force larger than their weight.
   c. objects submerged in water have a net force equal to their weight.
   d. objects submerged in water appear to weigh more than they do in air.

   ANS: A   DIF: I   OBJ: 9-1.4

8. Which of the following statements about floating objects is correct?
   a. The object's density is greater than the density of the fluid on which it floats.
   b. The object's density is equal to the density of the fluid on which it floats.
   c. The displaced volume of fluid is greater than the volume of the object.
   d. The buoyant force equals the object's weight.

   ANS: D   DIF: I   OBJ: 9-1.4

9. Which of the following statements about completely submerged objects resting on the ocean bottom is correct?
   a. The buoyant force acting on the object is equal to the object's weight.
   b. The apparent weight of the object depends on the object's density.
   c. The displaced volume of fluid is greater than the volume of the object.
   d. The weight of the object and the buoyant force are equal and opposite.

   ANS: B   DIF: I   OBJ: 9-1.4

10. A water bed that is 1.5 m wide and 2.5 m long weighs 1055 N. Assuming the entire lower surface of the bed is in contact with the floor, what is the pressure the bed exerts on the floor?
    a. 250 Pa   c. 270 Pa
    b. 260 Pa   d. 280 Pa

    ANS: D   DIF: IIIB   OBJ: 9-2.1

11. Each tire of an automobile has an area of 0.026 m^2 in contact with the ground. The weight of the automobile is 2.6×10^4 N. What is the pressure in the tires?
    a. 3.1×10^6 Pa   c. 2.5×10^5 Pa
    b. 6.5×10^3 Pa   d. 1.0×10^6 Pa

    ANS: C   DIF: IIIB   OBJ: 9-2.1

12. How much pressure is exerted on a swimmer at the bottom of a 5.0 m deep swimming pool?
    ($\rho_w = 1.00 \times 10^3$ kg/m^3, $P_0 = 1.01 \times 10^5$ Pa, and $g = 9.81$ m/s^2.)
    a. 1.5×10^5 Pa   c. 2.1×10^5 Pa
    b. 4.2×10^5 Pa   d. 1.0×10^6 Pa

    ANS: A   DIF: IIIB   OBJ: 9-2.2

13. What is the pressure at a depth of 580 m in a lake? ($\rho_w = 1.00 \times 10^3$ kg/m^3, $P_0 = 1.01 \times 10^5$ Pa, and $g = 9.81$ m/s^2)
    a. 3.7×10^6 Pa   c. 4.2×10^6 Pa
    b. 5.8×10^6 Pa   d. 8.5×10^6 Pa

    ANS: B   DIF: IIIB   OBJ: 9-2.2

14. How much pressure is exerted on a submarine at a depth of 8.50 km in the Pacific Ocean? (The density of sea water = 1.025×10^3 kg/m^3, and the atmospheric pressure at sea level = 1.01×10^5 Pa.)
    a. 8.6×10^5 Pa
    b. 9.5×10^6 Pa
    c. 8.7×10^6 Pa
    d. 8.6×10^6 Pa

    ANS: D   DIF: IIIC   OBJ: 9-2.2

15. The temperature in a container of fluid is
    a. a measure of the potential energy of the particles of the fluid.
    b. the total mass of the particles in the container.
    c. a measure of the average kinetic energy of the particles of the fluid.
    d. the number of particles in the container.

    ANS: C   DIF: I   OBJ: 9-2.3

16. Increasing the temperature of a fluid
    a. increases the speed of the particles.
    b. decreases the speed of the particles.
    c. decreases the number of particle collisions.
    d. decreases the pressure.

    ANS: A   DIF: I   OBJ: 9-2.3

17. What is 30°C on the Kelvin scale?
    a. −243 K
    b. 9 K
    c. 243 K
    d. 303 K

    ANS: D   DIF: I   OBJ: 9-2.3

18. What is 0°C on the Kelvin scale?
    a. −323 K
    b. −248 K
    c. 273 K
    d. 323 K

    ANS: C   DIF: I   OBJ: 9-2.3

19. An ideal fluid flows through a pipe made of two sections with diameters of 1.0 cm and 3.0 cm, respectively. By what factor would you have to multiply the velocity of the liquid flowing through the 1.0 cm section to obtain the velocity of liquid flowing through the 3.0 cm section?
    a. 6.0
    b. 9.0
    c. $\frac{1}{3}$
    d. $\frac{1}{9}$

    ANS: D   DIF: II   OBJ: 9-3.1

20. If the flow rate of a liquid is measured at 8.0×10^3 m^3/s going through a 2.0 cm radius pipe, which of the following is the average fluid velocity in the pipe?
    a. 0.64 m/s
    b. 2.0 m/s
    c. 4.0×10^{-2} m/s
    d. 6.4 m/s

    ANS: D   DIF: IIIB   OBJ: 9-3.1

21. The flow rate of blood through the average human aorta is about 9.0×10^1 cm³/s. If the aorta has a radius of 1.0 cm, what is the velocity of the blood flow?
   a. 14 cm/s
   b. 32 cm/s
   c. 37 cm/s
   d. 29 cm/s

   ANS: D   DIF: IIIB   OBJ: 9-3.1

22. A water tunnel has a circular cross section where the diameter diminishes from 3.6 m to 1.2 m. If the velocity of water flow is 3.0 m/s in the larger part of the tunnel, what is the velocity of flow in the smaller part of the tunnel?
   a. 9.0 m/s
   b. 18 m/s
   c. 27 m/s
   d. 54 m/s

   ANS: C   DIF: IIIA   OBJ: 9-3.1

23. Water flows into a house at a velocity of 15 m/s through a pipe that has a radius of 0.40 m. The water then flows through a smaller pipe at a velocity of 45 m/s. What is the radius of the smaller pipe?
   a. 0.53 m
   b. 0.23 m
   c. 0.17 m
   d. 0.37 m

   ANS: B   DIF: IIIB   OBJ: 9-3.1

24. In a machine, an ideal fluid with a density of 0.9×10^3 kg/m³ and a pressure of 1.3×10^5 Pa flows at a velocity of 6.0 m/s through a level tube with a radius of 0.5 cm. This tube connects to a level tube that has a radius of 1.5 cm. How fast does the water flow in the larger tube?
   a. 54 m/s
   b. 18 m/s
   c. 0.67 m/s
   d. 0.33 m/s

   ANS: C   DIF: IIIA   OBJ: 9-3.2

25. In a building, water flows at 15 m/s through a pipe that has a radius of 4.0×10^{-1} m. The water then rises 3.0 m to the second floor of the building. If the pressure remains unchanged, what is the speed of the water flow on the second floor?
   ($\rho_w = 1.00 \times 10^3$ kg/m³)
   a. 13 m/s
   b. 14 m/s
   c. 15 m/s
   d. 16 m/s

   ANS: A   DIF: IIIB   OBJ: 9-3.2

26. Water flows in a creek bed that has a diameter of 1.6 m at 25 m/s. In a tiny waterfall, the water drops 2.0 m. If the creek bed below has the same diameter, what would be the velocity of water flow after the waterfall? ($\rho_w = 1.00 \times 10^3$ kg/m³)
   a. 46 m/s
   b. 26 m/s
   c. 37 m/s
   d. 34 m/s

   ANS: B   DIF: IIIC   OBJ: 9-3.2

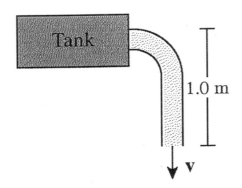

27. Gasoline is siphoned from a car tank, as shown in the figure above. The atmospheric pressure is the same at either end, and the height difference from the top of the tank to the bottom of the siphon is 1.0 m. Utilize Bernoulli's equation to determine the velocity of flow of gasoline out of the tube.
    a.   1.1 m/s
    b.   2.2 m/s
    c.   4.4 m/s
    d.   8.8 m/s

    ANS: C   DIF: IIIC   OBJ: 9-3.2

28. For an ideal fluid flowing through a horizontal pipe, Bernoulli's equation states that the sum of the pressure and energy per unit volume along the pipe does which of the following? (Assume measurements are taken along the pipe in the direction of fluid flow.)
    a.   increases as the pipe diameter increases
    b.   decreases as the pipe diameter increases
    c.   remains constant as the pipe diameter increases
    d.   increases, then decreases as the pipe diameter increases

    ANS: C   DIF: II   OBJ: 9-3.3

29. A municipal water supply is provided by a tall water tower. Water from this tower flows to a building. How does the water flow out of a faucet on the ground floor of a building compare with the water flow out of an identical faucet on the second floor of the building?
    a.   Water flows more rapidly out of the ground-floor faucet.
    b.   Water flows more rapidly out of the second-floor faucet.
    c.   Water flows at the same speed out of both faucets.
    d.   The speed of the water flow cannot be determined unless the height of the water tower is known.

    ANS: A   DIF: II   OBJ: 9-3.3

30. A person is standing near the edge of a railroad track when a high-speed train passes. The person tends to be
    a.   pushed away from the train.
    b.   pulled toward the train.
    c.   pushed upward into the air.
    d.   unaffected by the train.

    ANS: B   DIF: II   OBJ: 9-3.3

Holt Physics Assessment Item Listing
86

31. Two paper cups are hung by strings 10 cm apart. If a person blows between the cups,
    a. they move toward each other.   c. they do not move.
    b. they move away from each other.   d. they move upward.

    ANS: A   DIF: II   OBJ: 9-3.3

32. Which of the following is NOT a variable that describes the macroscopic state of an ideal gas?
    a. velocity   c. temperature
    b. pressure   d. volume

    ANS: A   DIF: I   OBJ: 9-4.1

33. At a constant volume, an ideal gas is heated from 348 K to 423 K. If the original pressure in the tank was 1.51×10^5 Pa and the volume does not change, what is the new pressure?
    a. 1.24×10^5 Pa   c. 1.84×10^5 Pa
    b. 9.75×10^4 Pa   d. 4.97×10^2 Pa

    ANS: C   DIF: IIIB   OBJ: 9-4.2

34. At a constant pressure, 6.00 m³ of an ideal gas at 348 K is cooled until its volume is halved. What is the new temperature of the gas?
    a. 174 K   c. 19.3 K
    b. 696 K   d. 116 K

    ANS: A   DIF: IIIA   OBJ: 9-4.2

35. The initial volume, pressure, and temperature of an ideal gas in a tank are 15 L, 2.0 atm, and 310 K, respectively. If the pressure increases to 3.5 atm and the temperature increases to 430 K, what is the final volume of the gas?
    a. 12 L   c. 3.1 L
    b. 6.2 L   d. 9.4 L

    ANS: A   DIF: IIIB   OBJ: 9-4.2

36. A container of helium gas is at a pressure of 6.0 atm and a temperature of 283 K. If the gas is heated at a constant volume until the pressure triples, what is the final temperature?
    a. 142 K   c. 849 K
    b. 94 K   d. 31 K

    ANS: C   DIF: IIIB   OBJ: 9-4.2

37. A cylinder with a movable piston contains an ideal gas at an initial temperature of 3.0×10^1 K, a volume of 1.5 m³, and a pressure of 0.20×10^5 Pa. What is the final volume of the gas if the temperature is increased to 287°C and the pressure remains constant?
    a. 3.0 m³   c. 1.0 m³
    b. 1.4 m³   d. 28 m³

    ANS: D   DIF: IIIB   OBJ: 9-4.2

SHORT ANSWER

1. Define a fluid.

   ANS:
   A fluid is a nonsolid state of matter in which the atoms or molecules are free to move past each other, as in a gas or liquid.

   DIF: I   OBJ: 9-1.1

2. What is the difference between a liquid and a gas?

   ANS:
   A liquid has a definite volume, and a gas does not.

   DIF: I   OBJ: 9-1.2

3. Does a gas change shape when it is poured from a small container to a large container? Explain.

   ANS:
   Yes. The gas expands and changes shape to fit the container.

   DIF: I   OBJ: 9-1.2

4. What is a liquid?

   ANS:
   A liquid is a fluid that has a definite volume but does not have a definite shape.

   DIF: I   OBJ: 9-1.2

5. What is a gas?

   ANS:
   A gas is a fluid that has neither a definite volume nor a definite shape.

   DIF: I   OBJ: 9-1.2

6. What determines whether an object will sink or float?

   ANS:
   The net force, or the apparent weight acting on the object, determines whether an object will sink or float.

   DIF: I   OBJ: 9-1.4

Holt Physics Assessment Item Listing
88

7. A water balloon weighing 19.6 N rests on a table. The balloon has an area of 0.015 m² in contact with the table. What pressure does the balloon exert on the table?

ANS:
1.31×10^3 Pa

DIF: IIIB   OBJ: 9.-2.1

8. What does the kinetic theory predict about temperature?

ANS:
Kinetic theory predicts that temperature is related to the average kinetic energy of the particles in a substance. As the temperature of a substance increases, the particles move faster, resulting in more particle collisions per unit of time.

DIF: I   OBJ: 9-2.3

9. Define *ideal gas*.

ANS:
An ideal gas is a gas whose behavior is accurately described by the ideal gas law, $PV = Nk_BT$.

DIF: I   OBJ: 9-4.1

10. Why does an ideal gas confined to a container NOT behave like an ideal fluid?

ANS:
An ideal gas confined to a container does not behave like an ideal fluid because it is no longer incompressible.

DIF: I   OBJ: 9-4.1

11. When does a real gas behave like an ideal gas?

ANS:
A real gas behaves like an ideal gas when it is at a relatively high temperature and at a relatively low pressure at room temperature and atmospheric pressure.

DIF: I   OBJ: 9-4.1

PROBLEM

1. An air-filled balloon with a mass of 3.0 g is placed in a pool of water. What is the buoyant force acting on the balloon if the density of air is 1.29×10^{-3} g/cm³ and the density of water is 1.00 g/cm³?

    ANS:
    2.9×10^{-2} N

    DIF: IIIA   OBJ: 9-1.3

2. An ice cube is placed in a glass of water. The cube is 2.0 cm on each side and has a density of 0.917 g/cm³. What is the buoyant force on the ice?

    ANS:
    7.2×10^{-2} N

    DIF: IIIA   OBJ: 9-1.3

MULTIPLE CHOICE

1. A substance's temperature increases as a direct result of
   a. energy being removed from the particles of the substance.
   b. kinetic energy being added to the particles of the substance.
   c. a change in the number of atoms and molecules in a substance.
   d. a decrease in the volume of the substance.

   ANS: B   DIF: I   OBJ: 10-1.1

2. What happens to the internal energy of an ideal gas when it is heated from 0°C to 4°C?
   a. It increases.   c. It remains constant.
   b. It decreases.   d. It is impossible to determine.

   ANS: A   DIF: II   OBJ: 10-1.1

3. Which of the following is proportional to the kinetic energy of atoms and molecules?
   a. elastic energy   c. potential energy
   b. temperature   d. thermal equilibrium

   ANS: B   DIF: I   OBJ: 10-1.1

4. Which of the following best describes the relationship between two systems in thermal equilibrium?
   a. No net energy is exchanged.   c. The masses are equal.
   b. The volumes are equal.   d. The velocity is zero.

   ANS: A   DIF: I   OBJ: 10-1.2

5. What is the temperature of a system in thermal equilibrium with another system made up of water and steam at 1 atm of pressure?
   a. 0°F   c. 0 K
   b. 273 K   d. 100°C

   ANS: D   DIF: II   OBJ: 10-1.2

6. What is the temperature of a system in thermal equilibrium with another system made up of ice and water at 1 atm of pressure?
   a. 0°F   c. 0 K
   b. 273 K   d. 100°C

   ANS: B   DIF: II   OBJ: 10-1.2

7. Heat flow occurs between two bodies in thermal contact when they differ in which of the following properties?
   a. mass   c. density
   b. specific heat   d. temperature

   ANS: D   DIF: I   OBJ: 10-2.1

8. If two small beakers of water, one at 70°C and one at 80°C, are emptied into a large beaker, what is the final temperature of the water?
   a. less than 70°C
   b. greater than 80°C
   c. between 70°C and 80°C
   d. The water temperature will fluctuate.

   ANS: C   DIF: II   OBJ: 10-1.2

9. All of the following are widely used temperature scales EXCEPT
   a. Kelvin.
   b. Fahrenheit.
   c. Celsius.
   d. Joule.

   ANS: D   DIF: I   OBJ: 10-1.3

10. If 546 K equals 273°C, then 500 K equals
    a. 227°C.
    b. 250°C.
    c. 773°C.
    d. 1000°C.

    ANS: A   DIF: IIIA   OBJ: 10-1.3

11. A substance registers a temperature change from 20°C to 40°C. This corresponds to an incremental change of
    a. 20°F.
    b. 40°F.
    c. 36°F.
    d. 313°F.

    ANS: C   DIF: IIIA   OBJ: 10-1.3

12. A substance registers a temperature change from 20°C to 40°C. This corresponds to an incremental change of
    a. 20 K.
    b. 40 K.
    c. 36 K.
    d. 313 K.

    ANS: A   DIF: IIIA   OBJ: 10-1.3

13. Which of the following is the equivalent of 88°F?
    a. 31°C
    b. 49°C
    c. 56°C
    d. 160°C

    ANS: A   DIF: IIIA   OBJ: 10-1.3

14. What temperature has the same numerical value on both the Celsius and the Fahrenheit scales?
    a. −40°
    b. 0°
    c. 40°
    d. −72°

    ANS: A   DIF: II   OBJ: 10-1.3

15. The average normal body temperature for human beings is 98.6°F. This corresponds to which of the following in degrees Kelvin?
    a. 296 K
    b. 310 K
    c. 393 K
    d. 273 K

    ANS: B   DIF: IIIA   OBJ: 10-1.3

16. If energy is transferred from a table to a block of ice moving across the table, which of the following statements is true?
    a. The table and the ice are at thermal equilibrium.
    b. The ice is cooler than the table.
    c. The ice is no longer 0°C.
    d. Energy is being transferred from the ice to the table.

    ANS: B   DIF: II   OBJ: 10-2.1

17. Why does sandpaper get hot when it is rubbed against rusty metal?
    a. Energy is transferred from the sandpaper into the metal.
    b. Energy is transferred from the metal to the sandpaper.
    c. Friction is creating the heat.
    d. Energy is transferred from a hand to the sandpaper.

    ANS: C   DIF: II   OBJ: 10-2.2

18. Energy transferred as heat always moves from an object
    a. at high temperature to an object at low temperature.
    b. at low temperature to an object at high temperature.
    c. at low kinetic energy to an object at high kinetic energy.
    d. of higher mass to an object of lower mass.

    ANS: A   DIF: I   OBJ: 10-2.1

19. Which of the following terms describes a transfer of energy?
    a. heat   c. temperature
    b. internal energy   d. kinetic energy

    ANS: A   DIF: I   OBJ: 10-2.1

20. If there is no temperature difference between a substance and its surroundings, what has occurred on the microscopic level?
    a. Energy was transferred from higher-energy particles to lower-energy particles.
    b. Energy was transferred from lower-energy particles to higher-energy particles.
    c. Thermal equilibrium was not reached.
    d. Heat has been flowing back and forth.

    ANS: A   DIF: II   OBJ: 10-2.2

21. High temperature is related to
    a. low kinetic energy.   c. no difference in kinetic energy.
    b. high kinetic energy.   d. zero net energy.

    ANS: B   DIF: I   OBJ: 10-2.2

22. A 5.00×10^2 kg object is attached by a rope through a pulley to a paddle-wheel shaft that is placed in a well-insulated tank holding 25.0 kg of water. The object is allowed to fall, causing the paddle wheel to rotate, churning the water. If the object falls a vertical distance of 1.00×10^2 m at constant speed, what is the temperature change of the water? ($c_p = 4186$ J/kg•°C and $g = 9.81$ m/s²)

a. 1.96×10^4°C
b. 4.69×10^3°C
c. 4.69°C
d. 0.800°C

ANS: C   DIF: IIIB   OBJ: 10-2.3

23. A 3.00×10^{-3} kg lead bullet travels at a speed of 2.40×10^2 m/s and hits a wooden post. If half the heat energy generated remains with the bullet, what is the increase in temperature of the embedded bullet? ($c_l = 1.28 \times 10^2$ J/kg•°C)

a. 112°C
b. 137°C
c. 225°C
d. 259°C

ANS: A   DIF: IIIB   OBJ: 10-3.1

24. What is the temperature increase of water per kilogram at the bottom of a 145 m waterfall if all of the initial potential energy is transferred as heat to the water? ($g = 9.81$ m/s² and $c_p = 4186$ J/kg•°C)

a. 0.170°C
b. 0.340°C
c. 0.680°C
d. 1.04°C

ANS: B   DIF: IIIB   OBJ: 10-2.3

25. What is the temperature increase of 4.0 kg of water when it is heated by an 8.0×10^2 W immersion heater for exactly 10.0 min? ($c_p = 4186$ J/kg•°C)

a. 57°C
b. 51°C
c. 29°C
d. 14°C

ANS: C   DIF: IIIB   OBJ: 10-2.3

26. A 0.2 kg mass of metal with a specific heat capacity of 1.26×10^3 J/kg•°C and an initial temperature of 90°C is placed in a 500 g calorimeter at an initial temperature of 20°C with a specific heat capacity of 4.19×10^2 J/kg•°C. The calorimeter is filled with 0.1 kg of water with an initial temperature of 20°C. When the combination of the metal, the calorimeter, and the water reaches equilibrium, what is the final temperature?

a. 70°C
b. 60°C
c. 50°C
d. 40°C

ANS: D   DIF: IIIC   OBJ: 10-3.1

27. Which of two rods has the greatest thermal conductivity?
    a. a rod with electrons that are freer to move from atom to atom than are the electrons another rod
    b. a rod with greater specific heat than another rod
    c. a rod with greater cross-sectional area than another rod
    d. a rod with greater length than another rod

    ANS: A   DIF: I   OBJ: 10-4.1

28. A 0.20 kg aluminum plate with an initial temperature of 20.0°C slides down a 15 m long surface that is inclined at a 30.0° angle to the horizontal. The force of kinetic friction exactly balances the component of gravity down the plane so that the plate, once started, glides down at constant velocity. If 90.0 percent of the mechanical energy of the system is absorbed by the aluminum, what is the temperature increase of the plate when it is at the bottom of the incline? ($c_a = 9.00 \times 10^2$ J/kg•°C)
    a. 0.16°C   c. 4.2×10^{-2}°C
    b. 7.2×10^{-2}°C   d. 3.1×10^{-2}°C

    ANS: B   DIF: IIIC   OBJ: 10-3.1

29. A 1.00×10^2 g piece of copper at an initial temperature of 95°C is dropped into 2.00×10^2 g of water contained in a 0.28 kg aluminum calorimeter. The water and calorimeter are initially at 15°C. What is the final temperature of the system when it reaches equilibrium? ($c_c = 3.9 \times 10^2$ J/kg•°C and $c_a = 9.00 \times 10^2$ J/kg•°C.)
    a. 16°C   c. 24°C
    b. 18°C   d. 25°C

    ANS: B   DIF: IIIC   OBJ: 10-3.1

30. A machine gear consists of 0.10 kg of iron and 0.16 kg of copper. How much total energy transfer as heat is generated in the gear if its temperature increases by 35°C? ($c_i = 4.6 \times 10^2$ J/kg•°C and $c_c = 3.9 \times 10^2$ J/kg•°C)
    a. 910 J   c. 5100 J
    b. 3800 J   d. 4400 J

    ANS: B   DIF: IIIC   OBJ: 10-3.1

31. An electric drill bores through a 0.100 kg piece of copper in 30.0 s. Find the increase in the temperature of the copper if the drill operates at 40.0 W. Assume that the drill does not increase in temperature. ($c_c = 387$ J/kg•°C)
    a. 40.6°C   c. 31.0°C
    b. 34.7°C   d. 27.3°C

    ANS: C   DIF: IIIB   OBJ: 10-3.1

32. Find the final equilibrium temperature when 10.0 g of milk at 10.0°C is added to 1.60×10^2 g of coffee with a temperature of 90.0°C. Assume the specific heats of coffee and milk are the same as for water ($c_w = 4.19$ J/g•°C), and disregard the heat capacity of the container.
    a. 85.3°C
    b. 77.7°C
    c. 71.4°C
    d. 66.7°C

    ANS: A   DIF: IIIB   OBJ: 10-3.1

33. A slice of bread contains about 4.19×10^5 J of energy. If the specific heat of a person is 4.19×10^3 J/kg•°C, by how many degrees Celsius would the temperature of a 70.0 kg person increase if all the energy in the bread were converted to heat?
    a. 2.25°C
    b. 1.86°C
    c. 1.43°C
    d. 1.00°C

    ANS: C   DIF: IIIB   OBJ: 10-3.1

34. A 3.0×10^{-3} kg lead bullet is traveling at a speed of 240 m/s when it becomes embedded in a block of ice with a temperature of 0.0°C. If all the heat generated goes into melting the ice, what quantity of ice is melted? ($L_f = 3.4 \times 10^5$ J/kg and $c_l = 1.3 \times 10^2$ J/kg•°C)
    a. 1.5×10^{-2} kg
    b. 5.8×10^{-4} kg
    c. 3.2×10^{-3} kg
    d. 2.5×10^{-4} kg

    ANS: D   DIF: IIIB   OBJ: 10-3.2

35. A flat container holds 200 g of water. Over a 10 min period, 1.5 g of water evaporates from the surface. What is the approximate temperature change of the remaining water? ($L_v = 2.26 \times 10^3$ J/g)
    a. 4°C
    b. –4°C
    c. 0.4°C
    d. –0.4°C

    ANS: B   DIF: IIIB   OBJ: 10-3.2

36. A pitcher of iced tea is made by adding ice to 1.8 kg of hot tea with an initial temperature of 80.0°C. How many kilograms of ice, which has an initial temperature of 0.0°C, are required to bring the mixture to 10.0°C? ($L_f = 3.3 \times 10^5$ J/kg)
    a. 1.8 kg
    b. 1.6 kg
    c. 1.4 kg
    d. 1.2 kg

    ANS: C   DIF: IIIB   OBJ: 10-3.2

37. A 5.0×10^2 g ice cube with an initial temperature of 0.0°C is placed in a plastic-foam box whose walls are 1.0 cm thick and whose total surface area is 6.0×10^2 cm^2. If the temperature of the air surrounding the box is exactly 20.0°C and it takes 4.0 h for the ice to completely melt, what is the conductivity of the plastic-foam material? ($L_f = 3.3 \times 10^2$ J/g)
    a. 4.0×10^{-2} J/m•s•°C
    b. 1.2×10^{-3} J/m•s•°C
    c. 5.0 J/m•s•°C
    d. 9.6×10^{-2} J/m•s•°C

    ANS: D   DIF: IIIC   OBJ: 10-4.1

38. A 1.0 kg cube of ice is dropped into 1.0 kg of water, and, when equilibrium is reached, there are 2.0 kg of ice at 0.0°C. The initial temperature of the water was 0°C. What was the original temperature of the ice? (c_w = 4186 J/kg•°C, c_i = 2093 J/kg•°C, and L_f= 3.3 × 10^5 J/kg•°C)

a. one or two degrees below 0.0°C
b. −80°C
c. −160°C
d. −240°C

ANS: C   DIF: IIIB   OBJ: 10-3.2

39. How much heat energy must be removed from 0.10 kg of oxygen with a temperature of 22.0°C in order for the oxygen to liquefy at −183°C? (c_0= 9.13 × 10^{-1} J/g•°C and L_v = 213 J/g.)

a. 5.71 × 10^4 J
b. 4.00 × 10^4 J
c. 1.81 × 10^4 J
d. 9.56 × 10^3 J

ANS: B   DIF: IIIB   OBJ: 10-3.2

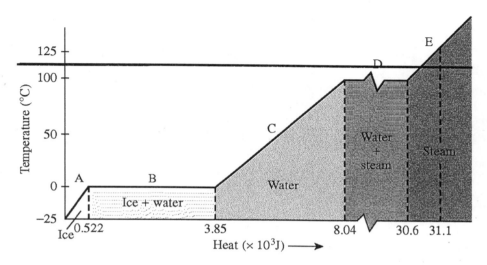

40. The figure above shows how the temperature of 10.0 g of ice changes as energy is added. Which of the following statements is correct?

a. The water absorbed energy continuously, but the temperature increased only when all of the water was in one phase.
b. The water absorbed energy sporadically, and the temperature increased only when all of the water was in one phase.
c. The water absorbed energy continuously, and the temperature increased continuously.
d. The water did not absorb energy.

ANS: A   DIF: II   OBJ: 10-3.3

41. At what point on the figure above is the amount of energy transferred as heat approximately 4.19 × 10^3 J?

a. A
b. B
c. C
d. D

ANS: C   DIF: II   OBJ: 10-3.3

42. At what point on the figure above does the substance undergo a phase change?
   a.   A   c.   C
   b.   B   d.   E

   ANS: A   DIF: II   OBJ: 10-3.3

43. Which of the following is a substance in which the temperature and pressure remain constant while the substance experiences an inward transfer of energy?
   a.   gas   c.   solid
   b.   liquid   d.   substance undergoing a change of state

   ANS: D   DIF: I   OBJ: 10-3.3

44. The use of fiberglass insulation in the outer walls of a building is intended to minimize heat transfer through what process?
   a.   conduction   c.   convection
   b.   radiation   d.   vaporization

   ANS: A   DIF: I   OBJ: 10-4.1

45. On a sunny day at the beach, the reason the sand gets hot and the water stays relatively cool is attributed to the difference in which property between water and sand?
   a.   mass density   c.   temperature
   b.   specific heat   d.   thermal conductivity

   ANS: B   DIF: I   OBJ: 10-4.1

SHORT ANSWER

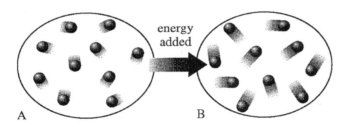

1. Describe how temperature is related to the kinetic energy of the molecules of the gas in the figure above.

   ANS:
   When energy is added to the gas, the kinetic energy of the molecules increases. The temperature increases because temperature is proportional to the kinetic energy of the molecules.

   DIF: II   OBJ: 10-1.1

2. Do "heat" and "cold" flow between objects? Explain.

   ANS:
   No, "heat" and "cold" do not flow between objects. Energy transferred between objects changes the temperature of the objects.

   DIF: II   OBJ: 10-2.1

3. Describe on the microscopic level why energy transfer as heat moves from an object at high temperature to an object at low temperature.

   ANS:
   An object at high temperature has higher-energy particles, and an object at low temperature has lower-energy particles. Energy is transferred as heat from the higher-energy particles to the lower-energy particles through collisions of the higher-energy particles with the lower-energy particles.

   DIF: II   OBJ: 10-2.2

4. On the microscopic level, explain the concept of heat transfer when a hand is placed in water that is 113°F.

   ANS:
   The molecules of water have a higher kinetic energy, so there are more particle collisions per unit of time per mass in the water than in the hand. Energy is transferred as heat to the outer layers of skin on the hand, which has fewer molecular collisions per unit of time per unit of mass.

   DIF: II   OBJ: 10-2.2

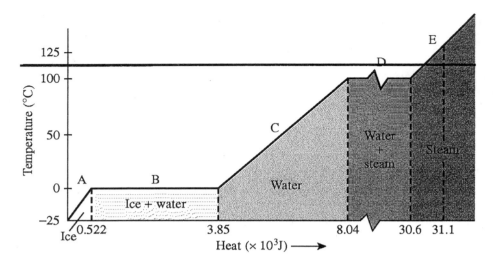

5. In the figure above, what happens to the ice at 0°C?

   ANS:
   The ice begins to melt and change into water.

   DIF: II   OBJ: 10-3.3

6. In the figure above, what happens to the ice at 100°C?

   ANS:
   The temperature stops rising, and the water turns into steam.

   DIF: II   OBJ: 10-3.3

7. What is a phase change?

   ANS:
   A phase change is the physical change of a substance from one state (solid, liquid, or gas) to another state at constant temperature and pressure.

   DIF: II   OBJ: 10-3.3

8. What is thermal conduction? What happens to atoms during thermal conduction?

   ANS:
   Thermal conduction is the process by which energy is transferred as heat through a material between two points at different temperatures. The atoms of the hotter substance lose energy, and the atoms of the cooler substance gain energy.

   DIF: I   OBJ: 10-4.1

9. What is hypothermia?

ANS:
Hypothermia is the transfer of too much energy from the human body to the surrounding air so that body temperature falls below normal levels.

DIF: I   OBJ: 10-4.2

10. Why is air an effective thermal insulator for the body?

ANS:
Air is an extremely poor thermal conductor. A thin layer of air near the skin provides a barrier from energy transfer.

DIF: I   OBJ: 10-4.2

11. Why would covering most of the body keep a person cool in the desert?

ANS:
Covering the body would protect it from direct sunlight and prevent excessive loss of body water through evaporation.

DIF: I   OBJ: 10-4.2

PROBLEM

1. The temperature of an oxygen tank is at 261 K, and the temperature of a nitrogen tank is 11°C. Which container has the higher temperature and by how many degrees?

ANS:
The nitrogen container is at the higher temperature by 23 K or 23°C.

DIF: IIIA   OBJ: 10-1.3

2. A mixture of 30.2 g of sand and 87.7 g of water has a temperature of 12.1°C. What mass of water at 85.8°C must be added to raise the final temperature of the mixture to 29.3°C? (c_w = 4.19 J/g•°C and c_s = 2.01 J/g•°C)

ANS:
31.1 g

DIF: IIIB   OBJ: 10-3.1

MULTIPLE CHOICE

1. An ideal gas is maintained at a constant pressure of 7.0×10^4 N/m² during an isobaric process while its volume decreases by 0.20 m³. What work is done by the system on its environment?
    a. 1.4×10^4 J
    b. 3.5×10^5 J
    c. -1.4×10^4 J
    d. -3.5×10^4 J

    ANS: C   DIF: IIIB   OBJ: 11-1.2

2. If an ideal gas does positive work on its surroundings,
    a. the temperature of the gas increases.
    b. the volume of the gas increases.
    c. the pressure of the gas increases.
    d. the internal energy of the gas decreases.

    ANS: B   DIF: II   OBJ: 11-1.2

3. A 2 mol ideal gas system is maintained at a constant volume of 4 L. If the pressure is constant, how much work is done by the system?
    a. 0 J
    b. 5 J
    c. 8 J
    d. 30 J

    ANS: A   DIF: II   OBJ: 11-1.2

4. A container of gas is at a pressure of 1.3×10^5 Pa and a volume of 6.0 m³. How much work is done by the gas if it expands at constant pressure to twice its initial volume?
    a. 3.7×10^5 J
    b. 2.6×10^5 J
    c. 7.8×10^5 J
    d. 4.6×10^5 J

    ANS: C   DIF: IIIB   OBJ: 11-1.2

5. A cylinder has a cross-sectional area of 0.02 m². How much work is done by a gas in the cylinder if the gas exerts a constant pressure of 7.8×10^5 Pa on the piston, moving it a distance of 0.06 m?
    a. 6.5×10^2 J
    b. 9.4×10^2 J
    c. 6.5×10^8 J
    d. 9.4×10^8 J

    ANS: B   DIF: IIIB   OBJ: 11-1.2

6. Air cools as it escapes from a diver's compressed air tank. What kind of process is this?
    a. isothermal
    b. isobaric
    c. adiabatic
    d. isovolumetric

    ANS: C   DIF: I   OBJ: 11-1.3

7. Sunlight strikes an ice cube at its melting point and causes it to melt. What is this process?
    a. adiabatic process
    b. isothermal process
    c. isobaric process
    d. isovolumetric process

    ANS: B   DIF: I   OBJ: 11-1.3

8. Which of the following is a thermodynamic process during which work is done on or by the system but no energy is transferred to or from the system as heat?
   a. adiabatic process
   b. isothermal process
   c. isovolumetric process
   d. isobaric process

   ANS: A   DIF: II   OBJ: 11-1.3

9. Which of the following processes for an ideal gas system has an unchanging internal energy and a heat intake that corresponds to the value of the work done by the system?
   a. isothermal process
   b. isobaric process
   c. adiabatic process
   d. isovolumetric process

   ANS: A   DIF: I   OBJ: 11-1.3

10. Which of the following is a thermodynamic process that takes place at a constant temperature and in which the internal energy of a system remains unchanged?
    a. adiabatic process
    b. isothermal process
    c. isovolumetric process
    d. isobaric process

    ANS: B   DIF: II   OBJ: 11-1.3

11. Which of the following is a thermodynamic process that takes place at constant volume so that no work is done on or by the system?
    a. adiabatic process
    b. isothermal process
    c. isovolumetric process
    d. isobaric process

    ANS: C   DIF: II   OBJ: 11-1.3

12. In an isovolumetric process by an ideal gas, the system's heat gain is equivalent to a change in which of the following?
    a. temperature
    b. volume
    c. pressure
    d. internal energy

    ANS: D   DIF: I   OBJ: 11-1.3

13. During an isovolumetric process, which of the following does not change?
    a. volume
    b. temperature
    c. internal energy
    d. pressure

    ANS: A   DIF: I   OBJ: 11-1.3

14. According to the first law of thermodynamics, the difference between energy transferred to or from a system as heat and energy transferred to or from a system by work is equivalent to which of the following?
    a. entropy change
    b. internal energy change
    c. temperature change
    d. specific heat

    ANS: B   DIF: I   OBJ: 11-2.1

15. How is conservation of internal energy expressed for an adiabatic system?
    a. $Q = W = 0$, so $\Delta U = 0$ and $U_i = U_f$
    b. $Q = 0$, so $\Delta U = -W$
    c. $\Delta T = 0$, so $\Delta U = 0$; therefore, $\Delta U = Q - W = 0$, or $Q = W$
    d. $\Delta V = 0$, so $P\Delta V = 0$ and $W = 0$; therefore, $\Delta U = Q$

    ANS: B   DIF: I   OBJ: 11-2.1

16. How is conservation of internal energy expressed for an isovolumetric system?
    a. $Q = W = 0$, so $\Delta U = 0$ and $U_i = U_f$
    b. $Q = 0$, so $\Delta U = -W$
    c. $\Delta T = 0$, so $\Delta U = 0$; therefore, $\Delta U = Q - W = 0$, or $Q = W$
    d. $\Delta V = 0$, so $P\Delta V = 0$ and $W = 0$; therefore, $\Delta U = Q$

    ANS: D   DIF: II   OBJ: 11-2.1

17. How is conservation of internal energy expressed for an isothermal system?
    a. $Q = W = 0$, so $\Delta U = 0$ and $U_i = U_f$
    b. $Q = 0$, so $\Delta U = -W$
    c. $\Delta T = 0$, so $\Delta U = 0$; therefore, $\Delta U = Q - W = 0$, or $Q = W$
    d. $\Delta V = 0$, so $P\Delta V = 0$ and $W = 0$; therefore, $\Delta U = Q$

    ANS: C   DIF: II   OBJ: 11-2.1

18. How is conservation of internal energy expressed for an isolated system?
    a. $Q = W = 0$, so $\Delta U = 0$ and $U_i = U_f$
    b. $Q = 0$, so $\Delta U = -W$
    c. $\Delta T = 0$, so $\Delta U = 0$; therefore, $\Delta U = Q - W = 0$, or $Q = W$
    d. $\Delta V = 0$, so $P\Delta V = 0$ and $W = 0$; therefore, $\Delta U = Q$

    ANS: A   DIF: II   OBJ: 11-2.1

19. A 4 mol ideal gas system undergoes an adiabatic process in which it expands and does 20 J of work on its environment. What is its change in internal energy?
    a. −20 J   c. 0 J
    b. −5 J   d. 20 J

    ANS: A   DIF: IIIA   OBJ: 11-2.2

20. A 4 mol ideal gas system undergoes an adiabatic process in which it expands and does 20 J of work on its environment. How much energy is transferred to the system as heat?
    a. −20 J   c. 5 J
    b. 0 J   d. 20 J

    ANS: B   DIF: IIIA   OBJ: 11-2.2

21. A total of 165 J of work is done on a gaseous refrigerant as it undergoes compression. If the internal energy of the gas increases by 123 J during the process, what is the total amount of energy transferred as heat?
   a. -42 J
   b. 42 J
   c. -288 J
   d. 288 J

   ANS: A   DIF: IIIA   OBJ: 11-2.2

22. The internal energy of a system is initially 63 J. A total of 71 J of energy is added to the system as heat while the system does 59 J of work. What is the system's final internal energy?
   a. 51 J
   b. 75 J
   c. 67 J
   d. 190 J

   ANS: B   DIF: IIIA   OBJ: 11-2.2

23. Over several cycles, a refrigerator does 1.73×10^4 J of work on the refrigerant. The refrigerant removes 8.11×10^4 J as heat from the air inside the refrigerator. How much energy is delivered to the outside air?
   a. 3.19×10^4 J
   b. 4.92×10^4 J
   c. 6.38×10^4 J
   d. 9.84×10^4 J

   ANS: D   DIF: IIIA   OBJ: 11-2.3

24. Over several cycles, a refrigerator does 5.13×10^4 J of work on the refrigerant. The refrigerant, in turn, removes 9.63×10^4 J as heat from the air inside the refrigerator. What is the net change in the internal energy of the refrigerant?
   a. 0.00 J
   b. 4.92×10^4 J
   c. 6.38×10^4 J
   d. 9.84×10^4 J

   ANS: A   DIF: II   OBJ: 11-2.3

25. Over several cycles, a refrigerator does 2.67×10^4 J of work on the refrigerant. The refrigerant, in turn, removes 7.49×10^4 J as heat from the air inside the refrigerator. How much work is done on the air inside the refrigerator?
   a. 0.00 J
   b. 4.92×10^4 J
   c. 6.38×10^4 J
   d. 9.84×10^4 J

   ANS: A   DIF: II   OBJ: 11-2.3

26. Over several cycles, a refrigerator does 1.73×10^4 J of work on the refrigerant. The refrigerant, in turn, removes 8.11×10^4 J as heat from the air inside the refrigerator. How much work is done on the air inside the refrigerator?
   a. 0.00 J
   b. 1.73×10^4 J
   c. -8.11×10^4 J
   d. 9.84×10^4 J

   ANS: C   DIF: II   OBJ: 11-2.3

27. An engine absorbs 2150 J as heat from a hot reservoir and gives off 750 J as heat to a cold reservoir during each cycle. How much work is done during each cycle?
   a. 750 J
   b. 1400 J
   c. 2150 J
   d. 2900 J

   ANS: B   DIF: II   OBJ: 11-2.3

28. According to the second law of thermodynamics, the heat received by a heat engine operating in a complete cycle from a high-temperature reservoir
   a. must be completely converted to work.
   b. equals the entropy increase.
   c. can be completely converted to internal energy.
   d. cannot be completely converted to work.

   ANS: D   DIF: II   OBJ: 11-3.1

29. The requirement that a heat engine must give up some energy at a lower temperature in order to do work corresponds to which law of thermodynamics?
   a. first
   b. second
   c. third
   d. No law of thermodynamics applies.

   ANS: B   DIF: I   OBJ: 11-3.1

30. A heat engine performs 2000.0 J of net work while adding 5000.0 J of heat to the cold-temperature reservoir. What is the efficiency of the engine?
   a. 0.714
   b. 0.600
   c. 0.400
   d. 0.286

   ANS: D   DIF: IIIB   OBJ: 11-3.2

31. An electrical power plant manages to transfer 88 percent of the heat produced in the burning of fossil fuel to convert water to steam. Of the heat carried by the steam, 40 percent is converted to the mechanical energy of the spinning turbine. Which best describes the overall efficiency of the heat-to-work conversion in the plant?
   a. greater than 88 percent
   b. 64 percent
   c. less than 40 percent
   d. 40 percent

   ANS: C   DIF: IIIA   OBJ: 11-3.2

32. A steam engine takes in 2.06×10^5 J of energy added as heat and exhausts 1.53×10^5 J of energy removed as heat per cycle. What is its efficiency?
   a. 0.743
   b. 0.346
   c. 0.257
   d. 0.673

   ANS: C   DIF: IIIB   OBJ: 11-3.2

33. A turbine exhausts 69 400 J of energy added as heat when it puts out 21 300 J of net work. What is the efficiency of the turbine?
   a. 3.26
   b. 0.307
   c. 0.693
   d. 0.235

   ANS: D   DIF: IIIB   OBJ: 11-3.2

34. An engine adds 75 000 J of energy as heat and removes 15 000 J of energy as heat. What is the engine's efficiency?
    a. 0.80
    b. 0.20
    c. 0.50
    d. 0.60

    ANS: A   DIF: IIIB   OBJ: 11-3.2

35. A ball is thrown against a brick wall. After the collision,
    a. the kinetic energy increases, and the ball is capable of doing more work.
    b. the kinetic energy decreases, and the ball is capable of doing less work.
    c. the kinetic energy increases, and the ball is capable of doing less work.
    d. the kinetic energy decreases, and the ball is capable of doing more work.

    ANS: B   DIF: II   OBJ: 11-4.1

36. When a system's disorder is increased,
    a. less energy is available to do work.
    b. more energy is available do work.
    c. no energy is available to do work.
    d. no work is done.

    ANS: A   DIF: I   OBJ: 11-4.1

37. Imagine you could observe the individual atoms that make up a piece of matter and that you observe the motion of the atoms becoming more orderly. What can you assume about the system?
    a. Its entropy is increasing.
    b. Its entropy is decreasing.
    c. It is gaining thermal energy.
    d. Positive work is being done on the system.

    ANS: A   DIF: I   OBJ: 11-4.2

38. A chunk of ice with a mass of 1 kg at 0°C melts and absorbs 3.35×10^8 J of heat in the process. Which best describes what happened to this system?
    a. Its entropy increased.
    b. Its entropy decreased.
    c. Its entropy remained constant.
    d. Work was converted to energy.

    ANS: A   DIF: I   OBJ: 11-4.2

39. According to the second law of thermodynamics, which of the following applies for any process that can occur within an isolated system?
    a. Entropy remains constant.
    b. Entropy increases.
    c. Entropy decreases.
    d. Entropy equals work done.

    ANS: B   DIF: I   OBJ: 11-4.2

40. When an egg is broken and scrambled, the entropy of the system
    a. increases, and the total entropy of the universe increases.
    b. decreases, and the total entropy of the universe increases.
    c. increases, and the total entropy of the universe decreases.
    d. decreases, and the total entropy of the universe decreases.

    ANS: A   DIF: II   OBJ: 11-4.3

41. When a drop of ink mixes with water, the entropy of the system
    a. increases, and the total entropy of the universe increases.
    b. decreases, and the total entropy of the universe increases.
    c. increases, and the total entropy of the universe decreases.
    d. decreases, and the total entropy of the universe decreases.

    ANS: A   DIF: II   OBJ: 11-4.3

42. When all the entropy changes in a process are included,
    a. the increases in entropy are always less than the decreases.
    b. the increases in entropy are always equal to the decreases.
    c. the increases in entropy are always greater than the decreases.
    d. the increases in entropy can be greater or less than the decreases.

    ANS: C   DIF: II   OBJ: 11-4.3

43. A thermodynamic process occurs, and the entropy of a system decreases. What can be concluded about the entropy change of the environment?
    a. It decreases.
    b. It increases.
    c. It stays the same.
    d. It could increase or decrease, depending on the process.

    ANS: B   DIF: II   OBJ: 11-4.3

SHORT ANSWER

1. A mechanic pushes down very quickly on the plunger of an insulated pump. The air hose is plugged so that no air escapes. Describe any transfer of energy as heat and any work done on or by the air in the system.

   ANS:
   No energy is transferred into or out of the system as heat. Work is done on the air in the system.

   DIF: II   OBJ: 11-1.1

2. A physics textbook is balanced on top of an inflated balloon on a cold morning. As the day passes, the temperature increases, the balloon expands, and the textbook rises. Is there a transfer of energy as heat? If so, what is it? Has any work been done?

   ANS:
   Energy from the air was transferred into the balloon as heat. The balloon did work on the book.

   DIF: II   OBJ: 11-1.1

3. A match is struck on a matchbook cover to create a flame. Is energy being transferred as heat to the system, or is work being done on the system?

ANS:
Work is being done on the system.

DIF: I   OBJ: 11-1.1

4. A system consists of a bomb, which explodes. Is energy being transferred as heat to the system, or is work being done on the system?

ANS:
Energy is being transferred as heat to the system.

DIF: I   OBJ: 11-1.1

5. What is true of the internal energy of an isolated system?

ANS:
According to the first law of thermodynamics, no energy is transferred to or from an isolated system.

DIF: I   OBJ: 11-2.1

6. How is the conservation of the internal energy of an isolated system expressed?

ANS:
It can be expressed as $Q = W = 0$ and as $\Delta U = Q - W = 0$.

DIF: I   OBJ: 11-2.1

7. According to the first law of thermodynamics, how can the internal energy of a system be increased?

ANS:
Energy can be transferred to the system as heat or by work.

DIF: I   OBJ: 11-2.1

8. State the second law of thermodynamics.

ANS:
No machine can be made that only absorbs energy as heat and then entirely transfers the energy out of the engine as an equal amount of work.

DIF: I   OBJ: 11-3.1

9. How does $Q_c > 0$ relate to the second law of thermodynamics?

ANS:
Some energy must always be transferred as heat to the system's surroundings.

DIF: I   OBJ: 11-3.1

10. How does using a screwdriver to insert a screw into a wood table increase entropy and reduce the ability of energy to do work?

ANS:
Instead of doing work on the screw, some of the energy goes toward increasing the internal energy of the screw and wood.

DIF: I   OBJ: 11-4.1

11. School office records are filed in folders in alphabetical order. Does this system have high or low entropy?

ANS:
low entropy

DIF: I   OBJ: 11-4.2

12. A backpack is stuffed with separate pages of notes from math, physics, and English classes. Does this system have high or low entropy?

ANS:
high entropy

DIF: I   OBJ: 11-4.2

13. A tray of cookies is removed from a box. Does this system have high or low entropy?

ANS:
low entropy

DIF: I   OBJ: 11-4.2

14. A glass beaker falls to the floor and breaks. Does this system have high or low entropy?

ANS:
high entropy

DIF: I   OBJ: 11-4.2

MULTIPLE CHOICE

1. Which of the following is NOT an example of approximate simple harmonic motion?
   a. a ball bouncing on the floor
   b. a child swinging on a swing
   c. a piano wire that has been struck
   d. a car's radio antenna waving back and forth

   ANS: A   DIF: I   OBJ: 12-1.1

2. Tripling the displacement from equilibrium of an object in simple harmonic motion will change the magnitude of the object's maximum acceleration by what factor?
   a. one-third   c. 3
   b. 1   d. 9

   ANS: C   DIF: I   OBJ: 12-1.2

3. A mass attached to a spring vibrates back and forth. At the equilibrium position, the
   a. the acceleration reaches a maximum.   c. net force reaches a maximum.
   b. velocity reaches a maximum.   d. velocity reaches zero.

   ANS: B   DIF: I   OBJ: 12-1.2

4. A mass attached to a spring vibrates back and forth. At maximum displacement, the spring force and the
   a. velocity reach a maximum.   c. acceleration reach a maximum.
   b. velocity reach zero.   d. acceleration reach zero.

   ANS: C   DIF: I   OBJ: 12-1.2

5. A simple pendulum swings in simple harmonic motion. At maximum displacement,
   a. the acceleration reaches a maximum.   c. the acceleration reaches zero.
   b. the velocity reaches a maximum.   d. the restoring forces reach zero.

   ANS: A   DIF: I   OBJ: 12-1.2

6. If a force of 50 N stretches a spring 0.10 m, what is the spring constant?
   a. 5 N/m   c. −5 N/m
   b. 500 N/m   d. −500 N/m

   ANS: B   DIF: IIIA   OBJ: 12-1.3

7. A 0.20 kg mass suspended from a spring moves with simple harmonic motion. At the instant the mass is displaced from equilibrium by −0.05 m, what is its acceleration? (The spring constant is 10.0 N/m.)
   a. 1200 m/s^2 c. 0.10 m/s^2
   b. 41 m/s^2 d. 2.5 m/s^2

   ANS: D   DIF: IIIA   OBJ: 12-1.3

8. How much displacement will a coil spring with a spring constant of 120 N/m achieve if it is stretched by a 60 N force?
   a. −0.5 m
   b. −2 m
   c. −4 m
   d. −7000 m

   ANS: A   DIF: IIIA   OBJ: 12-1.3

9. A mass on a spring that has been compressed 0.1 m has a restoring force of 20 N. What is the spring constant?
   a. 10 N/m
   b. 20 N/m
   c. 200 N/m
   d. 300 N/m

   ANS: C   DIF: IIIB   OBJ: 12-1.3

10. The angle between the string of a pendulum at its equilibrium position and at its maximum displacement is its
    a. period.
    b. frequency.
    c. vibration.
    d. amplitude.

    ANS: D   DIF: I   OBJ: 12-2.1

11. For a mass hanging from a spring, the maximum displacement the spring is stretched or compressed from its equilibrium position is its
    a. amplitude.
    b. period.
    c. frequency.
    d. acceleration.

    ANS: A   DIF: I   OBJ: 12-2.1

12. A pendulum swings through a total of 28°. If the displacement is equal on each side of the equilibrium position, what is the amplitude of this vibration? (Disregard frictional forces acting on the pendulum.)
    a. 28°
    b. 14°
    c. 56°
    d. 7.0°

    ANS: B   DIF: II   OBJ: 12-2.1

13. A child on a playground swings through a total of 32°. If the displacement is equal on each side of the equilibrium position, what is the amplitude of this vibration? (Disregard frictional forces acting on the swing.)
    a. 8.0°
    b. 16°
    c. 32°
    d. 64°

    ANS: B   DIF: II   OBJ: 12-2.1

14. Which of the following is the time it takes to complete a cycle of motion?
    a. amplitude
    b. period
    c. frequency
    d. revolution

    ANS: B   DIF: I   OBJ: 12-2.2

15. Which of the following is the number of cycles or vibrations per unit of time?
    a. amplitude
    b. period
    c. frequency
    d. revolution

    ANS: C   DIF: I   OBJ: 12-2.2

16. How are frequency and period related in simple harmonic motion?
    a. They are directly related.
    b. They are inversely related.
    c. They both measure the time per cycle.
    d. They both measure the number of cycles per unit of time.

    ANS: B   DIF: I   OBJ: 12-2.2

17. An amusement park ride has a frequency of 0.05 Hz. What is the ride's period?
    a. 5 s
    b. 10 s
    c. 20 s
    d. 40 s

    ANS: C   DIF: IIIA   OBJ: 12-2.3

18. Imagine that you could transport a simple pendulum from Earth to the moon, where the free-fall acceleration is one-sixth that on Earth. By what factor would the pendulum's frequency be changed?
    a. almost 6.0
    b. almost 2.5
    c. almost 0.4
    d. almost 0.17

    ANS: C   DIF: IIIB   OBJ: 12-2.3

19. An amusement park ride swings back and forth once every 40.0 s. What is the ride's frequency?
    a. 2.50×10^{-2} Hz
    b. 5.00×10^{-2} Hz
    c. 25.0×10^{-2} Hz
    d. 40.0×10^{-2} Hz

    ANS: A   DIF: IIIA   OBJ: 12-2.3

20. By what factor should the length of a simple pendulum be changed if the period of vibration were to be tripled?
    a. 3
    b. 6
    c. 9
    d. 27

    ANS: C   DIF: IIIB   OBJ: 12-2.3

21. A mass on a spring vibrates in simple harmonic motion at an amplitude of 8.0 cm. If the mass of the object is 0.20 kg and the spring constant is 130 N/m, what is the frequency?
    a. 1.5 Hz
    b. 8.7 Hz
    c. 4.0 Hz
    d. 1.6 Hz

    ANS: C   DIF: IIIB   OBJ: 12-2.3

22. A car with bad shock absorbers bounces up and down with a period of 1.5 s after hitting a bump. The car has a mass of 1500 kg and is supported by four springs with a spring constant of 6600 N/m. What is the period for each spring?
    a. 1.5 s   c. 4.4 s
    b. 5.8 s   d. 3.6 s

    ANS: A   DIF: IIIB   OBJ: 12-2.3

23. What is the period of a 4.12 m long pendulum?
    a. 2.01 s   c. 4.07 s
    b. 3.11 s   d. 9.69 s

    ANS: C   DIF: IIIB   OBJ: 12-2.3

24. On the planet Xenos, an astronaut observes that a 1.00 m long pendulum has a period of 1.50 s. What is the free-fall acceleration on Xenos?
    a. 4.18 m/s^2 c. 17.5 m/s^2
    b. 10.2 m/s^2 d. 26.3 m/s^2

    ANS: C   DIF: IIIB   OBJ: 12-2.3

25. Which of the following is a single nonperiodic disturbance?
    a. pulse wave   c. sine wave
    b. periodic wave   d. transverse wave

    ANS: A   DIF: I   OBJ: 12-3.2

26. Which of the following is a wave whose source is some form of repeating motion?
    a. pulse wave   c. sine wave
    b. periodic wave   d. transverse wave

    ANS: B   DIF: I   OBJ: 12-3.2

27. One end of a taut rope is fixed to a post. What type of wave is demonstrated if the free end is quickly raised and lowered?
    a. pulse wave   c. sine wave
    b. periodic wave   d. transverse wave

    ANS: A   DIF: I   OBJ: 12-3.2

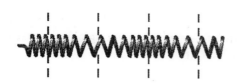

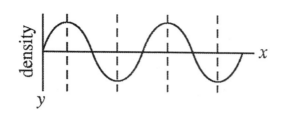

28. In the waveform of the longitudinal wave shown above, the compressed regions correspond to
    a. the wavelength.
    b. crests.
    c. troughs.
    d. the mass.

    ANS: B   DIF: I   OBJ: 12-3.3

29. In the waveform of the longitudinal wave shown above, the stretched regions correspond to
    a. the wavelength.
    b. crests.
    c. troughs.
    d. the mass.

    ANS: C   DIF: I   OBJ: 12-3.3

30. A periodic wave has a wavelength of 0.50 m and a speed of 20 m/s. What is the wave frequency?
    a. 0.02 Hz
    b. 20 Hz
    c. 40 Hz
    d. 10 Hz

    ANS: C   DIF: IIIA   OBJ: 12-3.4

31. A musical tone sounded on a piano has a frequency of 410 Hz and a wavelength of 0.80 m. What is the speed of the sound wave?
    a. 170 m/s
    b. 240 m/s
    c. 330 m/s
    d. 590 m/s

    ANS: C   DIF: IIIA   OBJ: 12-3.4

32. A radio wave has a speed of 3.00×10^8 m/s and a frequency of 107 MHz. What is the wavelength?
    a. 3.21 m
    b. 45.0 m
    c. 0.100 m
    d. 2.79 m

    ANS: D   DIF: IIIA   OBJ: 12-3.4

33. Bats can detect small objects, such as insects, that are approximately the size of one wavelength. If a bat emits a chirp at a frequency of 60.0 kHz and the speed of sound waves in air is 330 m/s, what is the size of the smallest insect that the bat can detect?
    a. 1.5 mm
    b. 3.5 mm
    c. 5.5 mm
    d. 7.5 mm

    ANS: C   DIF: IIIB   OBJ: 12-3.4

34. Waves propagate along a stretched string at a speed of 8.0 m/s. The end of the string vibrates up and down once every 1.5 s. What is the wavelength of the waves traveling along the string?

a. 3.0 m
b. 12 m
c. 6.0 m
d. 5.3 m

ANS: B   DIF: IIIB   OBJ: 12-3.4

35. What happens to the energy carried in a given time interval by a mechanical wave when the wave's amplitude is doubled?

a. It increases by a factor of two.
b. It increases by a factor of four.
c. It decreases by a factor of two.
d. It decreases by a factor of four.

ANS: B   DIF: I   OBJ: 12-3.5

36. What happens to the energy carried in a given time interval by a mechanical wave when the wave's amplitude is halved?

a. It increases by a factor of two.
b. It increases by a factor of four.
c. It decreases by a factor of two.
d. It decreases by a factor of four.

ANS: D   DIF: I   OBJ: 12-3.5

37. Two waves can occupy the same space at the same time because waves

a. are matter.
b. are displacements of matter.
c. do not cause interference patterns.
d. cannot pass through one another.

ANS: B   DIF: I   OBJ: 12-4.1

38. The superposition of mechanical waves can be observed in the movement of

a. bumper cars.
b. waves in a ripple tank.
c. electromagnetic radiation.
d. an orchestra.

ANS: B   DIF: I   OBJ: 12-4.1

39. What is the phase difference between two waves that are traveling in the same medium when they undergo constructive interference?

a. 270°
b. 180°
c. 90°
d. 0°

ANS: D   DIF: II   OBJ: 12-4.2

40. Which of the following is the interference that results when individual displacements on the same side of the equilibrium position are added together to form the resultant wave?

a. constructive
b. destructive
c. complete constructive
d. complete destructive

ANS: A   DIF: I   OBJ: 12-4.2

41. Which of the following is the interference that results when individual displacements on opposite sides of the equilibrium position are added together to form the resultant wave?

a. constructive
b. destructive
c. complete constructive
d. complete destructive

ANS: B   DIF: I   OBJ: 12-4.2

42. Which of the following types of interference will occur in the figure above?
    a. partial constructive
    b. partial destructive
    c. complete constructive
    d. complete destructive

    ANS: A   DIF: I   OBJ: 12-4.2

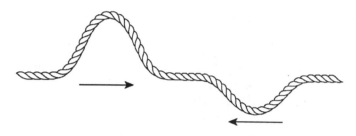

43. Which of the following types of interference will occur in the figure above?
    a. partial constructive
    b. partial destructive
    c. complete constructive
    d. complete destructive

    ANS: B   DIF: I   OBJ: 12-4.2

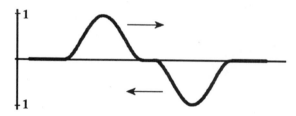

44. Which of the following types of interference will occur in the figure above?
    a. partial constructive
    b. partial destructive
    c. complete constructive
    d. complete destructive

    ANS: D   DIF: I   OBJ: 12-4.2

45. Consider two identical wave pulses on a rope. Suppose the first pulse reaches the fixed end of the rope, is reflected back, and then meets the second pulse. When the two pulses overlap exactly, the superposition principle predicts that the amplitude of the resultant pulses at that moment will be what factor times the amplitude of one of the original pulses?
    a. 0
    b. 1
    c. 2
    d. 4

    ANS: A   DIF: IIIA   OBJ: 12-4.3

46. At a fixed boundary, waves are
    a. neither reflected nor inverted.
    b. reflected but not inverted.
    c. reflected and inverted.
    d. inverted but not reflected.

    ANS: C   DIF: I   OBJ: 12-4.3

47. At a free boundary, waves are
    a. neither reflected nor inverted.
    b. reflected but not inverted.
    c. reflected and inverted.
    d. inverted but not reflected.

    ANS: B   DIF: I   OBJ: 12-4.3

48. A student sends a pulse traveling on a taut rope with one end attached to a post. What will the student observe?
    a. The pulse will not be reflected if the rope is free to slide up and down on the post.
    b. The pulse will be reflected and inverted if the rope is free to slide up and down on the post.
    c. The pulse will be reflected and inverted if the rope is fixed to the post.
    d. The pulse will not be inverted if the rope is fixed to the post.

    ANS: C   DIF: II   OBJ: 12-4.3

49. Standing waves are produced by periodic waves of
    a. any amplitude and wavelength traveling in the same direction.
    b. the same amplitude and wavelength traveling in the same direction.
    c. any amplitude and wavelength traveling in opposite directions.
    d. the same frequency, amplitude, and wavelength traveling in opposite directions.

    ANS: D   DIF: I   OBJ: 12-4.4

50. A 2.0 m long stretched rope is fixed at both ends. Which wavelength would NOT produce standing waves on this rope?
    a. 2.0 m
    b. 3.0 m
    c. 4.0 m
    d. 6.0 m

    ANS: B   DIF: IIIA   OBJ: 12-4.4

51. Which of the following wave lengths would produce standing waves on a string approximately 3.5 m long?
    a. 2.33 m
    b. 2.85 m
    c. 3.75 m
    d. 4.55 m

    ANS: A   DIF: IIIC   OBJ: 12-4.4

52. Which of the following wavelengths would NOT produce standing waves on a rope whose length is 1 m?
    a. 2/3 m
    b. 1 m
    c. 2 m
    d. π m

    ANS: D   DIF: I   OBJ: 12-4.4

53. How many nodes and antinodes are shown in the standing wave above?
   a. two nodes and three antinodes
   b. one node and two antinodes
   c. one-third node and one antinode
   d. three nodes and two antinodes

   ANS: D   DIF: I   OBJ: 12-4.5

54. A 3.0 m long stretched string is fixed at both ends. If standing waves with a wavelength of two-thirds L are produced on this string, how many nodes will be formed?
   a. 0
   b. 2
   c. 3
   d. 4

   ANS: D   DIF: II   OBJ: 12-4.5

55. What is the fewest number of nodes a standing wave can have?
   a. 1
   b. 2
   c. 3
   d. 4

   ANS: B   DIF: I   OBJ: 12-4.5

56. How many nodes and antinodes are shown in the standing wave above?
   a. four nodes and four antinodes
   b. four nodes and three antinodes
   c. four nodes and five antinodes
   d. five nodes and four antinodes

   ANS: D   DIF: I   OBJ: 12-4.5

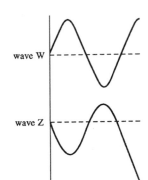

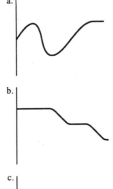

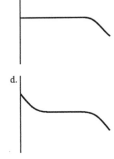

57. In the diagram above, use the superposition principle to find the resultant wave of waves **W** and **Z**.

    a. a   c. c
    b. b   d. d

    ANS: B   DIF: II   OBJ: 12.4.1

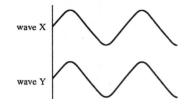

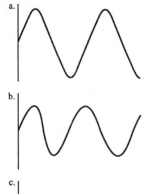

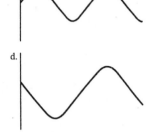

58. In the diagram above, use the superposition principle to find the resultant wave of waves X and Y.
    a. a
    b. b
    c. c
    d. d

    ANS: A   DIF: II   OBJ: 12.4.1

Holt Physics Assessment Item Listing
121

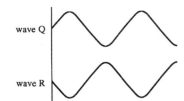

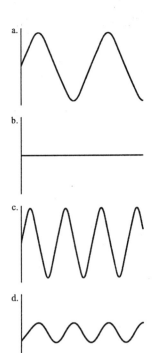

59. In the diagram above, use the superposition principle to find the resultant wave of waves Q and R.
   a. a
   b. b
   c. c
   d. d

   ANS: B

SHORT ANSWER

1. Is the motion of a metronome an example of simple harmonic motion?

   ANS:
   Yes

   DIF: I   OBJ: 12-1.1

2. Is the motion of a person dribbling a basketball an example of simple harmonic motion?

   ANS:
   No

   DIF: I   OBJ: 12-1.1

3. What is simple harmonic motion?

ANS:
Simple harmonic motion is vibration about an equilibrium position in which a restoring force is proportional to the displacement from equilibrium.

DIF: I   OBJ: 12-1.1

4. How is the relationship between period and frequency represented as an equation?

ANS:
$f = \dfrac{1}{T}$ or $T = \dfrac{1}{f}$

DIF: I   OBJ: 12-2.2

5. Suppose longitudinal simple harmonic waves are generated in a long spring. Describe the motion of a particle within the spring.

ANS:
The particle will vibrate continuously along the spring around an equilibrium position with simple harmonic motion.

DIF: II   OBJ: 12-3.1

6. Explain how particles in a medium are related to waves.

ANS:
A medium is a material through which a disturbance travels, and some waves cannot exist without it. The medium provides the particles that vibrate about an equilibrium position, that is, the particles through which a wave passes.

DIF: I   OBJ: 12-3.1

7. A boat produces a wave as it passes an aluminum can floating in a lake. Explain why the can is not carried by the wave motion.

ANS:
The water wave moves from one place to another, but the water is not carried with it. In other words, the disturbance moves from one location to another, but the medium does not.

DIF: I   OBJ: 12-3.1

8. Explain the relationship between local particle vibrations and overall wave motion.

ANS:
Waves are formed by dislocations of particles that are bound to equilibrium positions by restoring forces. As the particles move, they dislocate neighboring particles and the wave (disturbance) travels.

DIF: I   OBJ: 12-3.1

9. What is the difference between a pulse wave and a periodic wave?

ANS:
A pulse wave is a single traveling pulse, and a periodic wave is one whose source is a form of periodic motion.

DIF: I   OBJ: 12-3.2

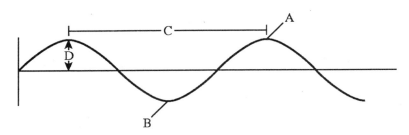

10. In the waveform shown above, which letter represents the amplitude of the wave?

ANS:
D

DIF: I   OBJ: 12-3.3

11. In the waveform shown above, which letter represents the trough of the wave?

ANS:
B

DIF: I   OBJ: 12-3.3

12. In the waveform shown above, what does letter C represent?

ANS:
wavelength

DIF: I   OBJ: 12-3.3

13. What happens to the energy of a wave when the amplitude is increased?

ANS:
The energy increases.

DIF: I   OBJ: 12-3.5

14. In a mechanical wave, what is the relationship between the energy and the wave's amplitude?

ANS:
The energy carried is proportional to the square of the wave's amplitude.

DIF: I   OBJ: 12-3.5

Holt Physics Assessment Item Listing
125

MULTIPLE CHOICE

1. Sound waves
   a. are a part of the electromagnetic spectrum.
   b. do not require a medium for transmission.
   c. are longitudinal waves.
   d. are transverse waves.

   ANS: C   DIF: I   OBJ: 13-1.1

2. The trough of the sine curve used to represent a sound wave corresponds to
   a. condensation.
   b. rarefaction.
   c. the point where molecules vibrate at a right angle to the direction of wave travel.
   d. a region of low elasticity.

   ANS: B   DIF: I   OBJ: 13-1.1

3. Which of the following is the region of a longitudinal wave in which the density and pressure are greater than normal?
   a. rarefaction   c. spherical wave
   b. compression   d. Doppler effect

   ANS: B   DIF: I   OBJ: 13-1.1

4. Which of the following is the region of a longitudinal wave in which the density and pressure are less than normal?
   a. rarefaction   c. spherical wave
   b. compression   d. Doppler effect

   ANS: A   DIF: I   OBJ: 13-1.1

5. How high or low we perceive a sound to be, depending on the frequency of the sound wave, is defined as the
   a. infrasonic wave.   c. ultrasonic wave.
   b. frequency.   d. pitch.

   ANS: D   DIF: I   OBJ: 13-1.2

6. Which of the following is the number of cycles per unit of time?
   a. infrasonic wave   c. ultrasonic wave
   b. frequency   d. pitch

   ANS: B   DIF: I   OBJ: 13-1.2

7. Pitch refers to
    a. how many vibrations per second are perceived by the human ear.
    b. how many cycles per second are in a transverse wave.
    c. the constructive interference of electromagnetic waves.
    d. the destructive interference in transverse waves.

    ANS: A   DIF: I   OBJ: 13-1.2

8. Which of the following has a higher speed of sound?
    a. helium at 0°C   c. copper at 0°C
    b. air at 0°C   d. air at 100°C

    ANS: C   DIF: I   OBJ: 13-1.3

9. Which has a higher speed of sound?
    a. water   c. methyl alcohol
    b. oxygen   d. copper

    ANS: D   DIF: I   OBJ: 13-1.3

10. Which statement about sound waves is correct?
    a. They generally travel faster through solids than through gases.
    b. They generally travel faster through gases than through solids.
    c. They generally travel faster through gases than liquids.
    d. They generally travel faster than light.

    ANS: A   DIF: I   OBJ: 13-1.3

11. Which of the following are series of compressions in graphical representations of spherical and plane waves?
    a. wavelength   c. rays
    b. source   d. wave fronts

    ANS: D   DIF: I   OBJ: 13-1.4

12. As a train starts from rest then accelerates down the track, coming toward an observer faster and faster, the frequency of the sound waves coming toward the observer will be
    a. less than the source frequency.   c. stationary.
    b. constantly increasing in frequency.   d. greater than the source frequency.

    ANS: B   DIF: II   OBJ: 13-1.5

13. The Doppler effect occurs with
    a. only sound waves.   c. only water waves.
    b. only compressional waves.   d. all waves.

    ANS: D   DIF: I   OBJ: 13-1.5

14. As a sound source approaches and then moves past a stationary observer, the observer will hear
    a. a steady rise in pitch.   c. a rise in pitch, then a drop in pitch.
    b. a sudden drop in pitch.   d. a drop in pitch, then a rise in pitch.

    ANS: C   DIF: I   OBJ: 13-1.5

15. If you hear the pitch of a siren become lower, you know that
    a. neither you nor the siren is moving.
    b. you are moving toward the siren or the siren is moving toward you.
    c. you are moving away from the siren or the siren is moving toward you.
    d. the source has just passed you or it is accelerating away from you.

    ANS: D   DIF: II   OBJ: 13-1.5

16. If you are on a train, how will the pitch of the train's whistle sound to you as the train moves?
    a. The pitch will become steadily higher.
    b. The pitch will become steadily lower.
    c. The pitch will not change.
    d. The pitch will become higher then become lower.

    ANS: C   DIF: II   OBJ: 13-1.5

17. The property of sound called intensity is
    a. proportional to the rate at which sound energy flows through an area normal to the direction of propagation.
    b. inversely proportional to the rate at which sound energy flows through an area normal to the direction of propagation.
    c. proportional to the period of a sound wave.
    d. proportional to the frequency of a sound wave.

    ANS: A   DIF: I   OBJ: 13-2.1

18. Tripling the distance from a sound source will change the intensity of the sound waves by what factor?
    a. $\frac{1}{9}$ c. 3

    b. $\frac{1}{3}$ d. 9

    ANS: A   DIF: II   OBJ: 13-2.1

19. What is the intensity of sound waves produced by a trumpet at a distance of 1.6 m when the power output of the trumpet is 0.30 W?
    a. 5.9×10^{-3} W/m^2 c. 9.4×10^{-3} W/m^2
    b. 1.5×10^{-2} W/m^2 d. 3.7×10^{-2} W/m^2

    ANS: C   DIF: IIIB   OBJ: 13-2.1

20. If the intensity of a sound is 8.0×10^{-4} W/m^2 at a distance of 5.0 m, what is the power of the sound?
    a. 1.2 W   c. 0.50 W
    b. 0.25 W   d. 1.6 W

    ANS: B   DIF: IIIB   OBJ: 13-2.1

21. If the intensity of a sound is increased by a factor of 100, the new decibel level will be
    a. two units greater.   c. 10 times greater.
    b. double the old one.   d. 20 units greater.

    ANS: D   DIF: IIIB   OBJ: 13-2.2

22. A difference in 10 dB means the sound is
    a. twice as loud.   c. 100 times as loud.
    b. 10 times as loud.   d. There would be no change in loudness.

    ANS: A   DIF: I   OBJ: 13-2.2

23. What is the intensity of a sound wave in relation to the intensity at the threshold of hearing?
    a. relative intensity   c. perceived loudness
    b. decibel level   d. resonance

    ANS: B   DIF: I   OBJ: 13-2.2

24. Which of the following is the condition that exists when the frequency of a force applied to a system matches the natural frequency of vibration of the system?
    a. pitch   c. timbre
    b. decibel level   d. resonance

    ANS: D   DIF: I   OBJ: 13-2.2

25. A sound twice the intensity of the faintest audible sound is not perceived as twice as loud because the sensation of loudness is
    a. approximately logarithmic in the human ear.
    b. approximately exponential in the human ear.
    c. outside the threshold of hearing.
    d. outside the threshold of pain.

    ANS: A   DIF: I   OBJ: 13-2.3

26. For a standing wave in an air column in a pipe that is open at both ends, there must be at least
    a. one node and one antinode.   c. two antinodes and one node.
    b. two nodes and one antinode.   d. two nodes and two antinodes.

    ANS: C   DIF: II   OBJ: 13-3.1

27. If both ends of a pipe are open,
    a. all harmonics are present.   c. only odd harmonics are present.
    b. no harmonics are present.   d. only even harmonics are present.

    ANS: A   DIF: I   OBJ: 13-3.1

28. If one end of a pipe is closed,
    a. all harmonics are present.   c. only odd harmonics are present.
    b. no harmonics are present.   d. only even harmonics are present.

    ANS: C   DIF: I   OBJ: 13-3.1

29. What is the lowest frequency that will resonate in a 2.0 m length organ pipe closed at one end? The speed of sound in air at this temperature is 340 m/s.
   a. 42 Hz
   b. 85 Hz
   c. 170 Hz
   d. 680 Hz

   ANS: A   DIF: IIIB   OBJ: 13-3.2

30. If a guitar string has a fundamental frequency of 500 Hz, what is the frequency of its second harmonic?
   a. 250 Hz
   b. 750 Hz
   c. 1000 Hz
   d. 1500 Hz

   ANS: C   DIF: IIIB   OBJ: 13-3.2

31. If a guitar string has a fundamental frequency of 7.50×10^2 Hz, what is the frequency of its fifth harmonic?
   a. 3750 Hz
   b. 750 Hz
   c. 2000 Hz
   d. 1500 Hz

   ANS: A   DIF: IIIB   OBJ: 13-3.2

32. The effects of sound on the ear are loudness, pitch, and quality. Loudness is an effect of _____, pitch is an effect of _____, and timbre is an effect of _____.
   a. intensity; harmonic content; frequency
   b. harmonic content; frequency; intensity
   c. frequency; intensity; harmonic content
   d. intensity; frequency; harmonic content

   ANS: D   DIF: I   OBJ: 13-3.3

33. Which of the following is the quality of a steady musical sound that is the result of a mixture of harmonics present at different intensities?
   a. tone
   b. timbre
   c. periodic waveform
   d. pitch

   ANS: B   DIF: I   OBJ: 13-3.3

34. Which describes an instrument's own characteristic full sound and mixture of harmonics at varying intensities?
   a. timbre
   b. harmonics
   c. pitch
   d. fundamental frequency

   ANS: A   DIF: I   OBJ: 13-3.3

35. What phenomenon is created by two tuning forks side by side that emit frequencies that differ by only a small amount?
   a. resonance
   b. interference
   c. the Doppler effect
   d. beats

   ANS: D   DIF: I   OBJ: 13-3.4

36. Two vibrating tuning forks held side by side will create a beat frequency of what value if the individual frequencies of the two forks are 342 Hz and 345 Hz, respectively?
   a. 687 Hz
   b. 343.5 Hz
   c. 339 Hz
   d. 3 Hz

   ANS: D   DIF: IIIA   OBJ: 13-3.4

37. Two vibrating tuning forks held side by side will create a beat frequency of what value if the individual frequencies of the two forks are 216 Hz and 224 Hz, respectively?
   a. 6 Hz
   b. 8 Hz
   c. 9 Hz
   d. 3 Hz

   ANS: B   DIF: IIIA   OBJ: 13-3.4

38. Two vibrating tuning forks held side by side will create a beat frequency of what value if the individual frequencies of the two forks are 398 Hz and 395 Hz, respectively?
   a. 6 Hz
   b. 5 Hz
   c. 9 Hz
   d. 3 Hz

   ANS: D   DIF: IIIA   OBJ: 13-3.4

39. Two vibrating tuning forks held side by side will create a beat frequency of what value if the individual frequencies of the two forks are 442 Hz and 449 Hz, respectively?
   a. 7 Hz
   b. 5 Hz
   c. 4 Hz
   d. 3 Hz

   ANS: A   DIF: IIIA   OBJ: 13-3.4

40. Two vibrating tuning forks held side by side will create a beat frequency of what value if the individual frequencies of the two forks are 512 Hz and 514 Hz, respectively?
   a. 2 Hz
   b. 5 Hz
   c. 9 Hz
   d. 6 Hz

   ANS: A   DIF: IIIA   OBJ: 13-3.4

41. Two vibrating tuning forks held side by side will create a beat frequency of what value if the individual frequencies of the two forks are 567 Hz and 565 Hz, respectively?
   a. 8 Hz
   b. 5 Hz
   c. 2 Hz
   d. 7 Hz

   ANS: C   DIF: IIIA   OBJ: 13-3.4

42. A vibrating guitar string emits a tone just as a 5.00×10^2 Hz tuning fork is struck. If a beat frequency of 5 Hz results, what is the possible frequency of vibration of the string?
   a. 2500 Hz
   b. 1500 Hz
   c. 605 Hz
   d. 495 Hz

   ANS: D   DIF: IIIA   OBJ: 13-3.4

43. Two notes have a beat frequency of 4 Hz. The frequency of one note is 420 Hz. What is the frequency of the other note?
    a. 422 Hz or 418 Hz
    b. 105 Hz
    c. 424 Hz or 416 Hz
    d. 1680 Hz

    ANS: C   DIF: IIIA   OBJ: 13-3.4

44. Two notes have a beat frequency of 6 Hz. The frequency of one note is 570 Hz. What is the frequency of the other note?
    a. 564 Hz or 576 Hz
    b. 5 Hz
    c. 424 Hz or 416 Hz
    d. 8 Hz

    ANS: A   DIF: IIIA   OBJ: 13-3.4

45. Two notes have a beat frequency of 8 Hz. The frequency of one note is 612 Hz. What is the frequency of the other note?
    a. 325 Hz or 318 Hz
    b. 5 Hz
    c. 604 Hz or 620 Hz
    d. 680 Hz

    ANS: C   DIF: IIIA   OBJ: 13-3.4

46. Two notes have a beat frequency of 10 Hz. The frequency of one note is 775 Hz. What is the frequency of the other note?
    a. 690 Hz or 700 Hz
    b. 765 Hz or 785 Hz
    c. 560 Hz or 580 Hz
    d. 325 Hz or 328 Hz

    ANS: B   DIF: IIIA   OBJ: 13-3.4

47. Two notes have a beat frequency of 3 Hz. The frequency of one note is 820 Hz. What is the frequency of the other note?
    a. 390 Hz or 393 Hz
    b. 855 Hz or 888 Hz
    c. 560 Hz or 580 Hz
    d. 817 Hz or 823 Hz

    ANS: D   DIF: IIIA   OBJ: 13-3.4

48. Two notes have a beat frequency of 5 Hz. The frequency of one note is 450 Hz. What is the frequency of the other note?
    a. 690 Hz or 700 Hz
    b. 567 Hz or 587 Hz
    c. 560 Hz or 580 Hz
    d. 445 Hz or 455 Hz

    ANS: D   DIF: IIIA   OBJ: 13-3.4

49. Beats are formed by the interference of two waves
    a. of slightly different frequencies traveling in the same direction.
    b. of slightly different frequencies traveling in different directions.
    c. with equal frequencies traveling in the same direction.
    d. with equal frequencies traveling in different directions.

    ANS: A   DIF: IIIA   OBJ: 13-3.4

50. Two violin players tuning their instruments together hear 8 beats in 2 s. What is the frequency difference between the two violins?
   a. 2 Hz
   b. 4 Hz
   c. 8 Hz
   d. 16 Hz

   ANS: B   DIF: II   OBJ: 13-3.4

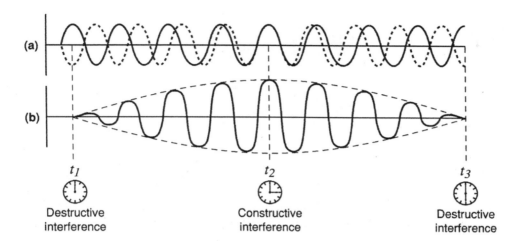

51. In the figure shown above, a beat occurs at
   a. t_1.
   b. t_2.
   c. t_3.
   d. t_1 and t_3.

   ANS: B   DIF: II   OBJ: 13-3.4

SHORT ANSWER

1. What determines the pitch of a musical note?

   ANS:
   frequency

   DIF: I   OBJ: 13-1.2

2. What happens to pitch when the frequency of a sound wave increases?

   ANS:
   The pitch rises.

   DIF: I   OBJ: 13-1.2

3. Which carries a sound wave more rapidly, a solid or a gas? Explain.

   ANS:
   A solid carries a sound wave more rapidly because its molecules are closer   together than those of a gas.

   DIF: I   OBJ: 13-1.3

4. How is a spherical wave graphically represented in two dimensions?

ANS:
Spherical waves are represented in two dimensions by a series of circles surrounding the source.

DIF: I   OBJ: 13-1.4

5. Can a spherical wave travel in one plane as a plane wave?

ANS:
yes

DIF: I   OBJ: 13-1.4

6. What are plane waves?

ANS:
As a sphere increases in radius, sections of its surface approach a plane surface. A plane wave is a section of a spherical wave that has such a large radius that sections of it appear planar. This condition appears when the observer of the wave is at a large distance from the source.

DIF: I   OBJ: 13-1.4

7. When does resonance occur?

ANS:
Resonance is produced when a forced vibration is the same as the natural frequency of an object.

DIF: I   OBJ: 13-2.3

8. Why would it be possible for an elevated section of roadway to collapse during an earthquake?

ANS:
If the earthquake waves have a frequency very close to the natural frequency of the section of roadway, the roadway will pick up the vibrations and collapse.

DIF: I   OBJ: 13-2.3

9. Why is the pattern of standing waves that occurs in a pipe open at both ends the same as that of a vibrating string?

ANS:
In both instances, there must be either a node or an antinode at each end but not one of each (as in a closed pipe).

DIF: I   OBJ: 13-3.1

10. How is it possible for some opera singers to shatter a crystal goblet with their voices?

    ANS:
    If an opera singer sings at a frequency very close to the natural frequency of the crystal goblet, the goblet will begin to vibrate and—if the motion builds up to a great enough amplitude—it will shatter.

    DIF: I   OBJ: 13-3.1

11. How is timbre related to harmonics?

    ANS:
    A mixture of harmonics account for timbre, or the sound quality of an instrument.

    DIF: I   OBJ: 13-3.3

PROBLEM

1. A resonating glass tube closed at one end is 4.0 cm wide and 47 cm long. What are the frequencies of the first three harmonics for the resonating tube? The speed of sound in air at this temperature is 346 m/s.

    ANS:
    180 Hz, 550 Hz, 920 Hz

    DIF: IIIB   OBJ: 13-3.2

MULTIPLE CHOICE

1. Which portion of the electromagnetic spectrum is used in a television?
   a. infrared waves
   b. microwaves
   c. radio waves
   d. gamma waves

   ANS: C   DIF: I   OBJ: 14-1.1

2. Which portion of the electromagnetic spectrum is used in aircraft navigation?
   a. infrared waves
   b. microwaves
   c. radio waves
   d. ultraviolet light

   ANS: B   DIF: I   OBJ: 14-1.1

3. Which portion of the electromagnetic spectrum is used in a microscope?
   a. infrared waves
   b. gamma rays
   c. visible light
   d. ultraviolet light

   ANS: C   DIF: I   OBJ: 14-1.1

4. Which portion of the electromagnetic spectrum is used to sterilize medical instruments?
   a. infrared waves
   b. microwaves
   c. X rays
   d. ultraviolet light

   ANS: D   DIF: I   OBJ: 14-1.1

5. Which portion of the electromagnetic spectrum is used to identify fluorescent minerals?
   a. ultraviolet light
   b. X rays
   c. infrared waves
   d. gamma rays

   ANS: A   DIF: I   OBJ: 14-1.1

6. What is the wavelength of microwaves of 3.0×10^9 Hz frequency?
   a. 0.060 m
   b. 0.10 m
   c. 0.050 m
   d. 0.20 m

   ANS: B   DIF: IIIB   OBJ: 14-1.2

7. What is the frequency of infrared light of 1.0×10^{-4} wavelength?
   a. 3.0×10^{-2} Hz
   b. 3.0×10^4 Hz
   c. 3.0×10^{12} Hz
   d. 3.0×10^2 Hz

   ANS: C   DIF: IIIB   OBJ: 14-1.2

8. What is the frequency of an electromagnetic wave with a wavelength of 1.0×10^5 m?
   a. 1.0×10^{13} Hz
   b. 3.0×10^3 Hz
   c. 3.0×10^{13} Hz
   d. 1.0×10^3 Hz

   ANS: B   DIF: IIIB   OBJ: 14-1.2

9. What is the wavelength of an infrared wave with a frequency of 4.2×10^{14} Hz?
   a. 7.1×10^5 m
   b. 1.4×10^6 m
   c. 7.1×10^{-7} m
   d. 1.4×10^{-6} m

   ANS: C   DIF: IIIB   OBJ: 14-1.2

10. Yellow-green light has wavelength of 560 m. What is its frequency?
    a. 5.4×10^6 Hz
    b. 1.8×10^6 Hz
    c. 1.8×10^{14} Hz
    d. 5.4×10^{14} Hz

    ANS: D   DIF: IIIB   OBJ: 14-1.2

11. In a vacuum, electromagnetic radiation of short wavelengths
    a. travels as fast as radiation of long wavelengths.
    b. travels slower than radiation of long wavelengths.
    c. travels faster than radiation of long wavelengths.
    d. can travel both faster and slower than radiation of long wavelengths.

    ANS: A   DIF: I   OBJ: 14-1.3

12. When red light is compared with violet light,
    a. both have the same frequency.
    b. both have the same wavelength.
    c. both travel at the same speed.
    d. red light travels faster than violet light.

    ANS: C   DIF: II   OBJ: 14-1.3

13. If you know the wavelength of any form of electromagnetic radiation, you can determine its frequency because
    a. all wavelengths travel at the same speed.
    b. the speed of light varies for each form.
    c. wavelength and frequency are equal.
    d. the speed of light increases as wavelength increases.

    ANS: A   DIF: I   OBJ: 14-1.3

14. The relationship between frequency, wavelength, and speed holds for light waves because
    a. light travels slower in a vacuum than in air.
    b. all forms of electromagnetic radiation travel at a single speed in a vacuum.
    c. light travels in straight lines.
    d. different forms of electromagnetic radiation travel at different speeds.

    ANS: B   DIF: II   OBJ: 14-1.3

15. The farther light is from a source,
    a. the more spread out light becomes.
    b. the more condensed light becomes.
    c. the more bright light becomes.
    d. the more light is available per unit area.

    ANS: A   DIF: I   OBJ: 14-1.4

16. If you are reading a book and you move twice as far away from the light source, how does the brightness at the new distance compare with that at the old distance? It is
    a. one-eighth
    b. one-fourth
    c. one-half
    d. twice

    ANS: B   DIF: II   OBJ: 14-1.4

17. Snow reflects almost all of the light incident upon it. However, a single beam of light is not reflected in the form of parallel rays. This is an example of _____ reflection off of a _____ surface.
    a. regular; rough
    b. regular; specular
    c. diffuse; specular
    d. diffuse; rough

    ANS: D   DIF: I   OBJ: 14-2.1

18. A highly polished finish on a new car provides a _____ surface for _____ reflection.
    a. rough; diffused
    b. specular; diffused
    c. rough; regular
    d. smooth; specular

    ANS: D   DIF: I   OBJ: 14-2.1

19. When incoming rays of light strike a flat mirror at an angle close to the surface of the mirror, the reflected rays are
    a. inclined high above the mirror's surface.
    b. parallel to the mirror's surface.
    c. perpendicular to the mirror's surface.
    d. close to the mirror's surface.

    ANS: D   DIF: I   OBJ: 14-2.2

20. When a straight line is drawn perpendicular to a flat mirror at the point where an incoming ray strikes the mirror's surface, the angles of incidence and reflection are measured from the normal and
    a. the angles of incidence and reflection are equal.
    b. the angle of incidence is greater than the angle of reflection.
    c. the angle of incidence is less than the angle of reflection.
    d. the angle of incidence can be greater than or less than the angle of reflection.

    ANS: A   DIF: I   OBJ: 14-2.2

21. If a light ray strikes a flat mirror at an angle of 27° from the normal, the reflected ray will be
    a. 27° from the mirror's surface.
    b. 27° from the normal.
    c. 90° from the mirror's surface.
    d. 63° from the normal.

    ANS: B   DIF: II   OBJ: 14-2.2

22. If a light ray strikes a flat mirror at an angle of 14° from the normal, the reflected ray will be
    a. 13° from the mirror's surface.
    b. 27° from the normal.
    c. 90° from the mirror's surface.
    d. 14° from the normal.

    ANS: D   DIF: II   OBJ: 14-2.2

Holt Physics Assessment Item Listing
138

23. If a light ray strikes a flat mirror at an angle of 29° from the normal, the reflected ray will be
    a. 29° from the normal.
    b. 27° from the normal.
    c. 90° from the mirror's surface.
    d. 36° from the normal.

    ANS: A   DIF: II   OBJ: 14-2.2

24. If a light ray strikes a flat mirror at an angle of 52° from the normal, the reflected ray will be
    a. 52° from the normal.
    b. 25° from the normal.
    c. 90° from the mirror's surface.
    d. 18° from the normal.

    ANS: A   DIF: II   OBJ: 14-2.2

25. If a light ray strikes a flat mirror at an angle of 61° from the normal, the reflected ray will be
    a. 61° from the mirror's surface.
    b. 27° from the normal.
    c. 90° from the mirror's surface.
    d. 61° from the normal.

    ANS: D   DIF: II   OBJ: 14-2.2

26. If a light ray strikes a flat mirror at an angle of 75° from the normal, the reflected ray will be
    a. 63° from the mirror's surface.
    b. 75° from the normal.
    c. 90° from the mirror's surface.
    d. 63° from the normal.

    ANS: B   DIF: II   OBJ: 14-2.2

27. If a light ray strikes a flat mirror at an angle of 30° from the normal, the ray will be reflected at an angle of
    a. 30° from the mirror's surface.
    b. 60° from the mirror's surface.
    c. 60° from the normal.
    d. 90° from the normal.

    ANS: B   DIF: IIIA   OBJ: 14-2.2

28. The image of an object in a flat mirror is always
    a. larger than the object.
    b. smaller than the object.
    c. independent of the size of the object.
    d. the same size as the object.

    ANS: D   OBJ: 14-2.3

29. When two parallel mirrors are placed so that their reflective sides face one another, _____ images form. This is because the image in one mirror becomes the _____ for the other mirror.
    a. multiple; object
    b. reduced; virtual image
    c. inverted; center of curvature
    d. enlarged; focal point

    ANS: A   DIF: I   OBJ: 14-2.3

30. If you stand 3.0 m in front of a flat mirror, how far away from you would your image be in the mirror?
    a. 1.5 m
    b. 3.0 m
    c. 6.0 m
    d. 12.0 m

    ANS: C   DIF: I   OBJ: 14-2.3

31. Which of the following best describes the image produced by a flat mirror?
    a. virtual, inverted, and magnification greater than one
    b. real, inverted, and magnification less than one
    c. virtual, upright, and magnification equal to one
    d. real, upright, and magnification equal to one

    ANS: C   DIF: I   OBJ: 14-2.3

32. When the reflection of an object is seen in a flat mirror, the distance from the mirror to the image depends on
    a. the wavelength of light used for viewing.
    b. the distance from the object to the mirror.
    c. the distance of both the observer and the object to the mirror.
    d. the size of the object.

    ANS: B   DIF: I   OBJ: 14-2.3

33. A concave mirror with a focal length of 10.0 cm creates a real image 30.0 cm away on its principal axis. How far from the mirror is the corresponding object?
    a. 20.0 cm   c. 7.50 cm
    b. 15.0 cm   d. 5.00 cm

    ANS: B   DIF: IIIB   OBJ: 14-3.1

34. A concave mirror forms a real image at 25 cm from the mirror surface along the principal axis. If the corresponding object is at a 10.0 cm distance, what is the mirror's focal length?
    a. 1.4 cm   c. 12 cm
    b. 17 cm   d. 7.1 cm

    ANS: D   DIF: IIIB   OBJ: 14-3.1

35. A concave mirror forms a real image at 14 cm from the mirror surface along the principal axis. If the corresponding object is at a 29 cm distance, what is the mirror's focal length?
    a. 14 cm   c. 12 cm
    b. 9.4 cm   d. 36 cm

    ANS: B   DIF: IIIB   OBJ: 14-3.1

36. A concave mirror forms a real image at 17 cm from the mirror surface along the principal axis. If the corresponding object is at a 36 cm distance, what is the mirror's focal length?
    a. 19 cm   c. 12 cm
    b. 47 cm   d. 26 cm

    ANS: C   DIF: IIIB   OBJ: 14-3.1

37. A concave mirror forms a real image at 26 cm from the mirror surface along the principal axis. If the corresponding object is at a 61 cm distance, what is the mirror's focal length?
    a. 12 cm   c. 22 cm
    b. 27 cm   d. 18 cm

    ANS: D   DIF: IIIB   OBJ: 14-3.1

38. A concave mirror forms a real image at 42 cm from the mirror surface along the principal axis. If the corresponding object is at a 88 cm distance, what is the mirror's focal length?
   a. 28 cm
   b. 17 cm
   c. 12 cm
   d. 9 cm

   ANS: A   DIF: IIIB   OBJ: 14-3.1

39. A concave mirror forms a real image at 19 cm from the mirror surface along the principal axis. If the corresponding object is at a 39 cm distance, what is the mirror's focal length?
   a. 13 cm
   b. 7 cm
   c. 11 cm
   d. 9 cm

   ANS: A   DIF: IIIB   OBJ: 14-3.1

40. If a virtual image is formed 10.0 cm along the principal axis from a convex mirror with a focal length of −15.0 cm, what is the object's distance from the mirror?
   a. 30.0 cm
   b. 12 cm
   c. 6.0 cm
   d. 3.0 cm

   ANS: A   DIF: IIIB   OBJ: 14-3.1

41. A convex mirror with a focal length of −20.0 cm forms an image 12 cm behind the surface. Where is the object as measured from the surface?
   a. 7.5 cm
   b. 15 cm
   c. 22 cm
   d. 3.0×10^1 cm

   ANS: D   DIF: IIIB   OBJ: 14-3.1

concave mirror

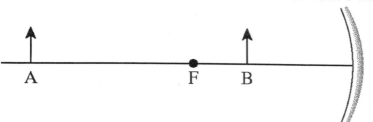

42. In the diagram above, the image of object *B* would be
   a. virtual, enlarged, and inverted.
   b. real, enlarged, and upright.
   c. virtual, reduced, and upright.
   d. virtual, enlarged, and upright.

   ANS: D   DIF: I   OBJ: 14-3.3

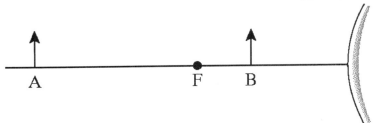

convex mirror

43. In the diagram above, the image of object *B* would be
    a. real, reduced, and upright.
    b. virtual, enlarged, and upright.
    c. virtual, reduced, and inverted.
    d. virtual, reduced, and upright.

    ANS: D   DIF: I   OBJ: 14-3.3

44. Which best describes the image of a concave mirror when the object is located somewhere between the focal point and twice the focal-point distance from the mirror?
    a. virtual, upright, and magnification greater than one
    b. real, inverted, and magnification less than one
    c. virtual, upright, and magnification less than one
    d. real, inverted, and magnification greater than one

    ANS: D   DIF: IIIB   OBJ: 14-3.3

45. Which best describes the image of a concave mirror when the object is at a distance greater than twice the focal-point distance from the mirror?
    a. virtual, upright, and magnification greater than one
    b. real, inverted, and magnification less than one
    c. virtual, upright, and magnification less than one
    d. real, inverted, and magnification greater than one

    ANS: B   DIF: IIIB   OBJ: 14-3.3

46. Which best describes the image of a concave mirror when the object's distance from the mirror is less than the focal-point distance?
    a. virtual, upright, and magnification greater than one
    b. real, inverted, and magnification less than one
    c. virtual, upright, and magnification less than one
    d. real, inverted, and magnification greater than one

    ANS: A   DIF: IIIB   OBJ: 14-3.3

47. When the reflection of an object is seen in a flat mirror, the image is
    a. real and upright.
    b. real and inverted.
    c. virtual and upright.
    d. virtual and inverted.

    ANS: C   DIF: I   OBJ: 14-3.3

48. When parallel rays that are also parallel to the principal axis strike a spherical mirror, rays that strike the mirror _____ the principal axis are focused at the focal point. Those rays that strike the mirror _____ the principal axis are focused at points between the mirror and the focal point.
   a. perpendicular to; far from
   b. close to; perpendicular to
   c. close to; far from
   d. far from; close to

   ANS: C   DIF: I   OBJ: 14-3.4

49. A parabolic mirror, instead of a spherical mirror, can be used to reduce the occurrence of which effect?
   a. spherical aberration
   b. mirages
   c. chromatic aberration
   d. light scattering

   ANS: A   DIF: I   OBJ: 14-3.4

50. When red light and green light shine on the same place on a piece of white paper, the spot appears to be
   a. yellow.
   b. brown.
   c. white.
   d. black.

   ANS: A   DIF: IIIB   OBJ: 14-4.1

51. Which of the following is NOT an additive primary color?
   a. yellow
   b. blue
   c. red
   d. green

   ANS: A   DIF: I   OBJ: 14-4.1

52. What color does yellow pigment subtract from white light?
   a. blue
   b. red
   c. yellow
   d. green

   ANS: A   DIF: I   OBJ: 14-4.2

53. What color does blue pigment subtract from white light?
   a. blue
   b. red
   c. violet
   d. green

   ANS: B   DIF: I   OBJ: 14-4.2

54. Which of the following is NOT a primary subtractive color?
   a. yellow
   b. cyan
   c. magenta
   d. blue

   ANS: D   DIF: I   OBJ: 14-4.2

55. A wave on a rope approaches two gratings in a row. The wave is polarized parallel to grating 1 and perpendicular to grating 2. The wave passes through
   a. only grating 1.
   b. only grating 2.
   c. both gratings.
   d. neither grating.

   ANS: A   DIF: II   OBJ: 14-4.3

56. As the angle is between the electric-field waves and the transmission axis increases,
    a. the component of light that passes through the polarizer decreases and the brightness of the light decreases.
    b. the component of light that passes through the polarizer decreases and the brightness of the light increases.
    c. the component of light that passes through the polarizer increases and the brightness of the light decreases.
    d. the component of light that passes through the polarizer increases and the brightness of the light increases.

    ANS: A   DIF: I   OBJ: 14-4.3

57. When the transmission axis is perpendicular to the plane of polarization for light,
    a. all the light passes through.   c. little of the light passes through.
    b. most of the light passes through.   d. no light passes through.

    ANS: D   DIF: I   OBJ: 14-4.3

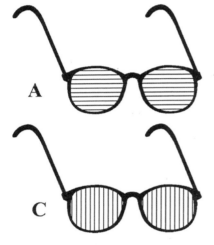
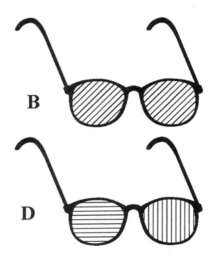

58. Which pair of glasses is best suited for automobile drivers? The transmission axes are shown by straight lines on the lenses. (Hint: The light reflects off the hood of the car.)
    a. A   c. C
    b. B   d. D

    ANS: C   DIF: II   OBJ: 14-4.3

59. If you looked at a light through the lenses from two polarizing sunglasses that were overlapped at right angles to one another,
    a. all of the light would pass through.   c. little of the light would pass through.
    b. most of the light would pass through.   d. none of the light would pass through.

    ANS: D   DIF: II   OBJ: 14-4.3

SHORT ANSWER

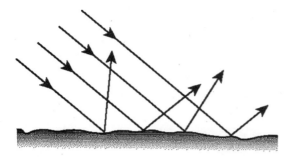

1. What type of reflection is illustrated in the figure above?

    ANS:
    diffuse

    DIF: I   OBJ: 14-2.1

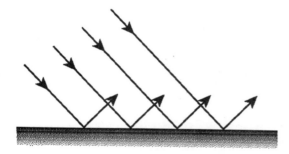

2. What type of reflection is illustrated in the figure above?

    ANS:
    specular

    DIF: I   OBJ: 14-2.1

3. When rays of light are incident upon a spherical mirror far from the principal axis, fuzzy images form. What is this characteristic of spherical mirrors?

    ANS:
    spherical aberration

    DIF: I   OBJ: 14-3.4

4. Spherical aberration may be avoided by employing a _____ mirror or by making sure that the diameter of a spherical mirror is sufficiently _____.

    ANS:
    parabolic; small

    DIF: I   OBJ: 14-3.4

Holt Physics Assessment Item Listing
145

5. The focal point and center of curvature of a spherical mirror all lie along the
_____.

ANS:
principal axis

DIF: I   OBJ: 14-3.4

6. Why are some primary colors called additive?

ANS:
When they are added in varying proportions, they can form all of the colors of the spectrum.

DIF: I   OBJ: 14-4.1

7. What occurs when beams of light of three primary colors are combined?

ANS:
They form white light.

DIF: I   OBJ: 14-4.1

8. What occurs when light passed through a red filter is combined with light passed through a green filter?

ANS:
Yellow light appears.

DIF: I   OBJ: 14-4.1

9. What occurs when yellow light is combined with blue light?

ANS:
White or colorless light appears.

DIF: I   OBJ: 14-4.1

10. What occurs when subtractive primary colors are combined?

ANS:
When they are combined, they filter out all light.

DIF: I   OBJ: 14-4.2

11. What color results when yellow and blue pigment are combined?

ANS:
green

DIF: I   OBJ: 14-4.2

12. How is the brightness of a light source affected by distance?

ANS:
Brightness decreases by the square of the distance from the source.

DIF: II   OBJ: 14-1.4

PROBLEM

1. A certain radio wave has a frequency of 2.0×10^6 Hz. What is its wavelength?

ANS:
150 m

DIF: IIIB   OBJ: 14-1.2

2. Where would the image of a 4.0 cm tall object that is 12 cm in front of a flat mirror be located?

ANS:
12 cm directly behind the mirror

DIF: I   OBJ: 14-2.3

3. A convex mirror has a focal length of −17 cm. What is the radius of curvature?

ANS:
−34 cm

DIF: IIIB   OBJ: 14-3.1

4. A candle is 49.0 cm in front of a convex spherical mirror with a radius of curvature of 70.0 cm. Draw a ray diagram to find the position and magnification of the image.

ANS:
$q = -20.4$ cm; $M = +0.417$

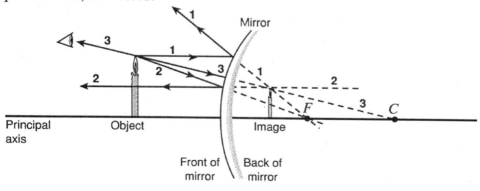

DIF: IIIB   OBJ: 14-3.2

5. An object that is 2.00 cm high is placed 10.0 cm in front of a concave mirror with a radius of curvature of 40.0 cm. Draw a ray diagram to find the position and magnification of the image.

ANS:
$q = +20.0$ cm; $M = +2.00$

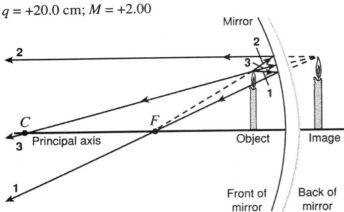

DIF: IIIB   OBJ: 14-3.2

6. An object is 15 cm from the surface of a spherical glass tree ornament that is 5.0 cm in diameter. Draw a ray diagram to find the position and magnification of the image.

ANS:
$q = -1.15$ cm; $M = 7.7 \times 10^{-2}$

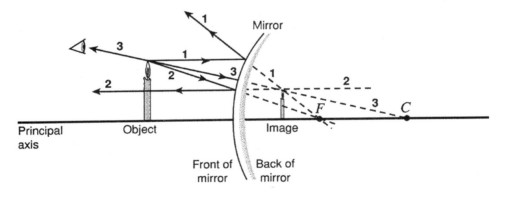

DIF: IIIB   OBJ: 14-3.2

7. A concave spherical mirror has a radius of curvature of 10.0 cm. A candle that is 5.0 cm tall is placed 14 cm in front of the mirror. Draw a ray diagram to find the image distance and magnification.

ANS:
$q = 7.8$ cm; $M = -0.56$

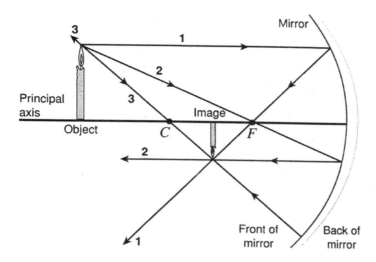

DIF: IIIB   OBJ: 14-3.2

8. A candle 4.7 cm tall is 20.0 cm from a convex mirror that has a focal length of 6.0 cm. Draw a ray diagram to determine the position and magnification of the image.

ANS:
$q = -4.6$ cm; $M = +0.23$

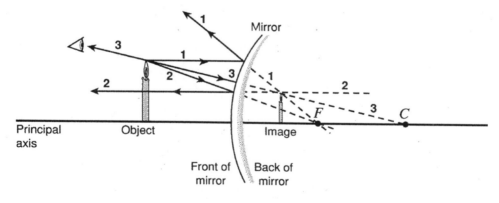

DIF: IIIB   OBJ: 14-3.2

9. A candle 15 cm high is placed in front of a concave mirror at the focal point. The radius of curvature is 60 cm. Draw a ray diagram to determine the position and magnification of the image.

ANS:
When the candle is at the focal point, the image is infinitely far to the left and therefore is not seen, as shown in the answer diagram.

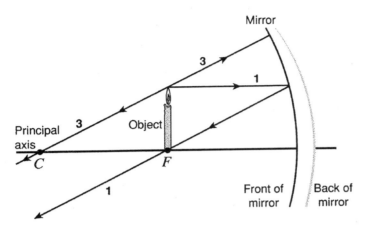

DIF: I   OBJ: 14-3.2

MULTIPLE CHOICE

1. Part of a pencil that is placed in a glass of water appears bent in relation to the part of the pencil that extends out of the water. What is this phenomenon called?
   a. interference
   b. refraction
   c. diffraction
   d. reflection

   ANS: B   DIF: I   OBJ: 15-1.1

2. Refraction is the term for the bending of a wave disturbance as it passes at an angle from one _____ into another.
   a. glass
   b. medium
   c. area
   d. boundary

   ANS: B   DIF: I   OBJ: 15-1.1

3. Which is an example of refraction?
   a. A parabolic mirror in a headlight focuses light into a beam.
   b. A fish appears closer to the surface of the water than it really is when observed from a riverbank.
   c. In a mirror, when you lift your right arm, the left arm of your image is raised.
   d. Light is bent slightly around corners.

   ANS: B   DIF: II   OBJ: 15-1.1

4. The _____ of light can change when light is refracted because the wavelength changes.
   a. frequency
   b. media
   c. color
   d. transparency

   ANS: C   DIF: I   OBJ: 15-1.1

5. Light is NOT refracted when it is
   a. traveling from air into a glass of water at an angle of 35° to the normal.
   b. traveling from water into air at an angle of 35° to the normal.
   c. striking a wood surface.
   d. traveling from air into a diamond at an angle of 45°.

   ANS: C   DIF: I   OBJ: 15-1.1

6. When light passes at an angle to the normal from one material into another material in which its speed is higher,
   a. it is bent toward the normal to the surface.
   b. it always lies along the normal to the surface.
   c. it is unaffected.
   d. it is bent away from the normal to the surface.

   ANS: D   DIF: I   OBJ: 15-1.2

7. When light passes at an angle to the normal from one material into another material in which its speed is lower,
   a. it is bent toward the normal to the surface.
   b. it always lies along the normal to the surface.
   c. it is unaffected.
   d. it is bent away from the normal to the surface.

   ANS: A   DIF: I   OBJ: 15-1.2

8. When a light ray moves from air into glass at an angle of 45°, its path is
   a. bent toward the normal.   c. parallel to the normal.
   b. bent away from the normal.   d. not bent.

   ANS: A   DIF: II   OBJ: 15-1.2

9. When a light ray passes from water ($n = 1.333$) into diamond ($n = 2.419$) at an angle of 45°, its path is
   a. bent toward the normal.   c. parallel to the normal.
   b. bent away from the normal.   d. not bent.

   ANS: A   DIF: II   OBJ: 15-1.2

10. When a light ray passes from zircon ($n = 1.923$) into fluorite ($n = 1.434$) at an angle of 60°, its path is
    a. bent toward the normal.   c. parallel to the normal.
    b. bent away from the normal.   d. not bent.

    ANS: B   DIF: II   OBJ: 15-1.2

11. A ray of light in air is incident on an air-to-glass boundary at an angle of exactly 3.0×10^{1}° with the normal. If the index of refraction of the glass is 1.65, what is the angle of the refracted ray within the glass with respect to the normal?
    a. 56°   c. 29°
    b. 46°   d. 18°

    ANS: D   DIF: IIIB   OBJ: 15-1.3

12. Carbon tetrachloride ($n = 1.46$) is poured into a container made of crown glass ($n = 1.52$). If a light ray in the glass is incident on the glass-to-liquid boundary and makes an angle of 30.0° with the normal, what is the angle of the corresponding refracted ray with respect to the normal?
    a. 55.5°   c. 31.4°
    b. 28.7°   d. 19.2°

    ANS: C   DIF: IIIB   OBJ: 15-1.3

13. A lapidary cuts a diamond so that the light will refract at an angle of 17.0° to the normal. What is the index of refraction of the diamond when the angle of incidence is 45.0°?
    a. 2.13   c. 1.23
    b. 1.74   d. 2.42

    ANS: D   DIF: IIIB   OBJ: 15-1.3

14. A beam of light in air is incident at an angle of 35° to the surface of a rectangular block of clear plastic ($n = 1.49$). What is the angle of refraction?
    a. 42°   c. 55°
    b. 23°   d. 59°

    ANS: B   DIF: IIIB   OBJ: 15-1.3

15. An object is placed along the principal axis of a thin converging lens that has a focal length of 30.0 cm. If the distance from the object to the lens is 40.0 cm, what is the distance from the image to the lens?
    a. 17.3 cm   c. 1.20 m
    b. −17.3 cm   d. −1.20 m

    ANS: C   DIF: II   OBJ: 15-2.2

16. An object is placed 30.0 cm from a thin converging lens along the axis of the lens. The lens has a focal length of 10.0 cm. What is the distance from the image to the lens?
    a. 15.0 cm   c. 60.0 cm
    b. −15.0 cm   d. −60.0 cm

    ANS: A   DIF: IIIB   OBJ: 15-2.2

17. An object is placed 6.0 cm from a thin converging lens along the axis of the lens. The lens has a focal length of 9.0 cm. What is the distance from the image to the lens?
    a. 3.0 cm   c. 18 cm
    b. −3.0 cm   d. −18 cm

    ANS: D   DIF: IIIB   OBJ: 15-2.2

18. An object is placed 20.0 cm from a thin converging lens along the axis of the lens. If a real image forms behind the lens at a distance of 8.00 cm from the lens, what is the focal length of the lens?
    a. 30.0 cm   c. 15.0 cm
    b. 10.0 cm   d. 5.71 cm

    ANS: A   DIF: IIIB   OBJ: 15-2.2

19. An object is placed 14.0 cm from a diverging lens. If a virtual image appears 10.0 cm from the lens on the same side as the object, what is the focal length of the lens?
    a. −50.0 cm   c. −10.0 cm
    b. −35 cm   d. −8.0 cm

    ANS: B   DIF: IIIB   OBJ: 15-2.2

20. An object is placed 40.0 cm from a converging lens along the axis of the lens. If a virtual image forms at a distance of 50.0 cm from the lens on the same side as the object, what is the focal length of the lens?
    a. 22.0 cm   c. 90.0 cm
    b. 45.0 cm   d. 2.00 m

    ANS: D   DIF: IIIB   OBJ: 15-2.2

21. A film projector produces a 1.51 m image of a horse on a screen. If the projector lens is 4.00 m from the screen and the size of the horse on the film is 1.07 cm, what is the magnification of the image?
    a. 141
    b. −14.1
    c. 7.08×10^{-3}
    d. -7.08×10^{-3}

    ANS: A   DIF: IIIB   OBJ: 15-2.3

22. A candle that is 10.0 cm high is placed 30.0 cm in front of a diverging lens. The lens has a focal length of −20.0 cm. What is the magnification of the image?
    a. 2.50
    b. −0.400
    c. 0.400
    d. −2.50

    ANS: C   DIF: IIIB   OBJ: 15-2.3

23. An object that is 18 cm from a converging lens forms a real image 22.5 cm from the lens. What is the magnification of the image?
    a. −1.25
    b. 1.25
    c. −0.80
    d. 0.80

    ANS: A   DIF: IIIB   OBJ: 15-2.3

24. A converging lens has a focal length of 10.0 cm. If a virtual image of an object is formed 25.0 cm in front of the lens, what is the magnification of the image?
    a. −3.50
    b. 3.50
    c. 1.50
    d. −1.50

    ANS: B   DIF: IIIC   OBJ: 15-2.3

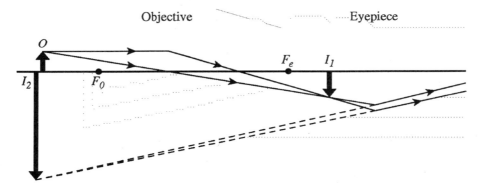

25. In the diagram of a compound microscope above, where would you place the slide?
    a. at O
    b. at I_2
    c. at F_0
    d. at I_1

    ANS: A   DIF: I   OBJ: 15-2.4

26. Which of the following describes what will happen to a light ray incident on a glass-to-air boundary at greater than the critical angle?
    a. total reflection
    b. total transmission
    c. partial reflection, partial transmission
    d. partial reflection, total transmission

    ANS: A   DIF: I   OBJ: 15-3.1

27. Atmospheric refraction of light rays is responsible for which effect?
    a. spherical aberration
    c. chromatic aberration
    b. mirages
    d. total internal reflection in a gemstone

    ANS: B   DIF: IIIB   OBJ: 15-3.2

28. If atmospheric refraction did not occur, how would the apparent time of sunrise and sunset be changed?
    a. Both would be later.
    b. Both would be earlier.
    c. Sunrise would be later, and sunset would be earlier.
    d. Sunrise would be earlier, and sunset would be later.

    ANS: C   DIF: I   OBJ: 15-3.2

29. Which is NOT correct when describing the formation of rainbows?
    a. A rainbow is really spherical in nature.
    b. Sunlight is spread into a spectrum when it enters a spherical raindrop.
    c. Sunlight is internally reflected on the back side of a raindrop.
    d. The angle between the incident white light and the returning violet ray is 45°.

    ANS: D   DIF: I   OBJ: 15-3.3

SHORT ANSWER

1. When does refraction occur?

    ANS:
    Refraction occurs when light's velocity changes.

    DIF: I   OBJ: 15-1.1

2. How are two converging lenses used to view an object in a compound microscope?

    ANS:
    An object placed just outside the focal length of the objective lens forms a real, inverted image just inside the focal point of the eyepiece. This eyepiece, the second lens, serves to magnify the image.

    DIF: I   OBJ: 15-2.4

3. Why is it impossible to see an atom with a compound microscope?

    ANS:
    In order to be seen, the object under a microscope must be at least as large as a wavelength of light. An atom is several times smaller than a wavelength of visible light.

    DIF: II   OBJ: 15-2.4

4. In a refracting telescope, why does the eyepiece act as a simple magnifier?

    ANS:
    It acts as a simple magnifier because it is positioned so that its focal point lies very close to the focal point of the objective lens.

    DIF: I   OBJ: 15-2.4

5. In a refracting telescope, is the image upright or inverted? Explain.

    ANS:
    It is inverted on the objective lens, and the eyepiece magnifies this inverted object.

    DIF: I   OBJ: 15-2.4

6. The critical angle for internal reflection inside a certain transparent material is found to be 48°. If entering light has an angle of incidence of 52°, predict whether the light will be refracted or whether it will undergo total internal reflection.

    ANS:
    The light will undergo total internal reflection.

    DIF: IIIB   OBJ: 15-3.1

7. A ray of light travels from calcite ($n = 1.434$) into air at an angle of 35°. Predict whether the light will be refracted or whether it will undergo total internal reflection.

    ANS:
    The light will be refracted.

    DIF: IIIB   OBJ: 15-3.1

8. At the end of the day, why are we able to see the sun after it has passed below the horizon?

    ANS:
    Rays of light from the sun strike Earth's atmosphere and are bent because the atmosphere has an index of refraction different from that of the near-vacuum of space.

    DIF: I   OBJ: 15-3.2

9. What atmospheric conditions produce a mirage?

    ANS:
    A mirage may be produced when the ground is so hot that the air above the ground is warmer than air at higher altitudes. The difference in temperature causes the air to have different indexes of refraction, so refraction occurs. This sends light from the sky upward into an observer's eyes.

    DIF: I   OBJ: 15-3.2

Holt Physics Assessment Item Listing
156

10. Why do motorists sometimes see what appear to be wet spots on the road on a dry summer day?

ANS:
Light rays from the blue sky above are refracted by the warm air next to the dark, hot road and end up traveling upward into the motorists' eyes.

DIF: I   OBJ: 15-3.2

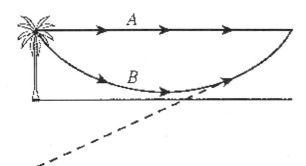

11. Use the figure above to describe how a mirage is produced.

ANS:
In this situation, the observer sees the palm tree in two different ways. One group of light rays reaches the observer by the straight-line path *A*. In addition, a second group of rays travels along the curved path *B* because of refraction. Consequently, the observer also sees an inverted image of the palm tree.

DIF: I   OBJ: 15-3.2

12. What is dispersion?

ANS:
Dispersion is the process of separating polychromatic light into its component wavelengths because n is a function of wavelength for all material mediums. Snell's law says that the angles of refraction will be different for different wavelengths even if the angles of incidence are the same.

DIF: I   OBJ: 15-3.2

13. How does white light passing through a prism produce a visible spectrum?

ANS:
Each colored component of the incoming ray is refracted depending on its wavelength. The rays fan out from the second face of the prism to produce a visible spectrum.

DIF: I   OBJ: 15-3.2

Holt Physics Assessment Item Listing
157

14. What does the perceived color of each water droplet in a rainbow depend on?

ANS:
The perceived color depends on the angle at which that drop is viewed.

DIF: I   OBJ: 15-3.2

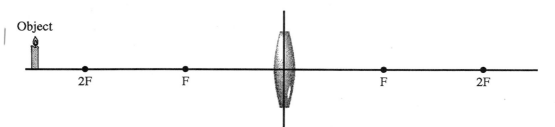

15. What is the position and kind of image produced by the lens above? Draw a ray diagram to support your answer.

ANS:
A real image will be produced between *F* and 2*F*.

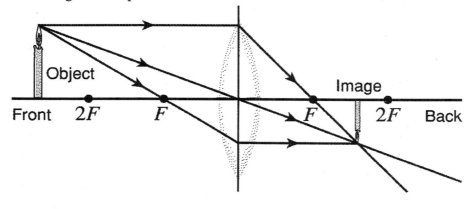

DIF: II   OBJ: 15-2.1

16. A student burns a hole in a pencil with a magnifying lens. What is the position and kind of image produced by the lens? Draw a ray diagram to support your answer.

ANS:
A virtual image will be produced at *F*.

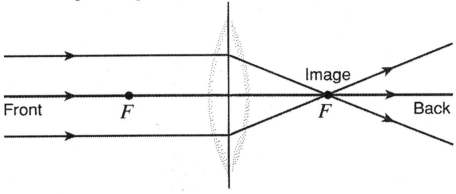

DIF: II   OBJ: 15-2.1

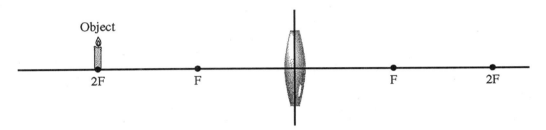

17. What is the position and kind of image produced by the lens above? Draw a ray diagram to support your answer.

ANS:
A real image will be produced at 2*F*.

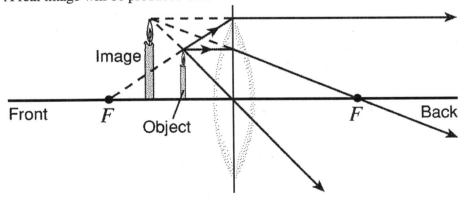

DIF: II   OBJ: 15-2.1

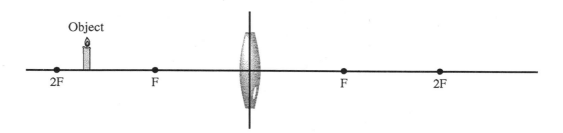

18. What is the position and kind of image produced by the lens above? Draw a ray diagram to support your answer.

ANS:
A real image will be produced outside 2F.

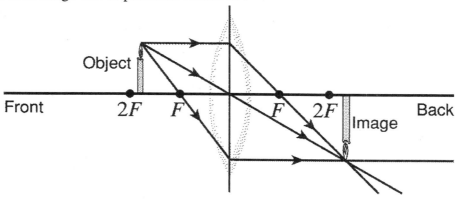

DIF: II   OBJ: 15-2.1

PROBLEM

1. A ray of light passes from air into carbon disulfide ($n = 1.63$) at an angle of 28.0° to the normal. What is the refracted angle?

ANS:
16.7°

DIF: IIIA   OBJ: 15-1.3

2. A ray of light passes from air into cubic zirconia ($n = 2.20$) at an angle of 56° to the normal. What is the refracted angle?

ANS:
122°

DIF: IIIA   OBJ: 15-1.3

3. A ray of light passes from air into ice ($n = 1.309$) at an angle of 46° to the normal. What is the refracted angle?

ANS:
33°

DIF: IIIA   OBJ: 15-1.3

4. A ray of light passes from air into glycerine ($n = 1.473$) at an angle of 29° to the normal. What is the refracted angle?

ANS:
19°

DIF: IIIA   OBJ: 15-1.3

5. A ray of light passes from air into water ($n = 1.333$) at an angle of 55° to the normal. What is the refracted angle?

ANS:
38°

DIF: IIIA   OBJ: 15-1.3

6. A ray of light passes from air into fluorite ($n = 1.434$) at an angle of 19° to the normal. What is the refracted angle?

ANS:
13°

DIF: IIIA   OBJ: 15-1.3

7. An object is placed along the principal axis of a thin converging lens that has a focal length of 16 cm. If the distance from the object to the lens is 24 cm, what is the distance from the image to the lens?

ANS:
0.48 m

DIF: IIIB   OBJ: 15-2.2

8. An object is placed along the principal axis of a thin converging lens that has a focal length of 22 cm. If the distance from the object to the lens is 36 cm, what is the distance from the image to the lens?

ANS:
0.57 m

DIF: IIIB   OBJ: 15-2.2

9. An object is placed along the principal axis of a thin converging lens that has a focal length of 46 cm. If the distance from the object to the lens is 63 cm, what is the distance from the image to the lens?

ANS:
170 cm

DIF: IIIB   OBJ: 15-2.2

10. An object is placed along the principal axis of a thin converging lens that has a focal length of 39 cm. If the distance from the object to the lens is 51 cm, what is the distance from the image to the lens?

ANS:
160 cm

DIF: IIIB   OBJ: 15-2.2

11. An object is placed along the principal axis of a thin converging lens that has a focal length of 21.0 cm. If the distance from the object to the lens is 35.0 cm, what is the distance from the image to the lens?

ANS:
52.5 cm

DIF: IIIB   OBJ: 15-2.2

12. An object is placed along the principal axis of a thin converging lens that has a focal length of 19 cm. If the distance from the object to the lens is 27 cm, what is the distance from the image to the lens?

ANS:
64 cm

DIF: IIIB   OBJ: 15-2.2

13. A ray of light travels across a liquid-to-glass interface. The index of refraction for the liquid is 1.75 and 1.52 for the glass. If the light meets the interface at an angle of 59°, predict whether the light will be refracted or whether it will undergo total internal reflection.

ANS:
The light will be refracted.

DIF: IIIB   OBJ: 15-3.1

Holt Physics Assessment Item Listing
162

14. A fiber-optic cable ($n = 1.50$) is submerged in water ($n = 1.33$). Predict whether light will be refracted or whether it will undergo total internal reflection if the angle of incidence is between 45° and 50°.

ANS:
The light will undergo total internal reflection.

DIF: IIIB   OBJ: 15-3.1

15. An optical fiber is made of clear plastic ($n = 1.50$). Light travels through the fiber at angles ranging from 43° to 59°. Predict whether the light will be refracted or whether it will undergo total internal reflection when the cable is in the air.

ANS:
There will be total internal reflection.

DIF: IIIB   OBJ: 15-3.1

MULTIPLE CHOICE

1. In a double-slit interference pattern the path length from one slit to the first dark fringe of a double-slit interference pattern is longer than the path length from the other slit to the fringe by
    a. three-quarters of a wavelength.
    c. one-quarter of a wavelength.
    b. one-half of a wavelength.
    d. one full wavelength.

    ANS: B   DIF: I   OBJ: 16-1.1

2. Interference effects observed in the early 1800s were instrumental in supporting a concept of the existence of which property of light?
    a. polarization
    c. wave nature
    b. particle nature
    d. electromagnetic character

    ANS: C   DIF: I   OBJ: 16-1.1

3. In Young's double-slit experiment, a wave from one slit arrives at a point on a screen one wavelength behind the wave from the other slit. What is observed at that point?
    a. dark fringe
    c. multicolored fringe
    b. bright fringe
    d. gray fringe, neither dark nor bright

    ANS: B   DIF: I   OBJ: 16-1.1

4. In order to produce a sustained interference pattern by light waves from multiple sources, which condition or conditions must be met?
    a. Sources must be coherent.
    b. Sources must be monochromatic.
    c. Sources must be coherent and monochromatic.
    d. Sources must be neither coherent nor monochromatic.

    ANS: C   DIF: I   OBJ: 16-1.2

5. Two beams of coherent light are shining on the same sheet of white paper. When referring to the crests and troughs of such waves, where will darkness appear on the paper?
    a. where the crest from one wave overlaps the crest from the other
    b. where the crest from one wave overlaps the trough from the other
    c. where the troughs from both waves overlap
    d. Darkness cannot occur because the two waves are coherent.

    ANS: B   DIF: I   OBJ: 16-1.2

6. For stable interference to occur, the phase difference must be
    a. $\dfrac{V}{C}$.
    c. $\frac{1}{2}\lambda$.
    b. $c\lambda$.
    d. constant.

    ANS: D   DIF: II   OBJ: 16-1.2

7. The distance between the two slits in a double-slit experiment is 0.04 mm. The second-order bright fringe ($m = 2$) is measured on a screen at an angle of 2.2° from the central maximum. What is the wavelength of the light?

   a. 560 nm
   b. 630 nm
   c. 750 nm
   d. 770 nm

   ANS: D   DIF: IIIB   OBJ: 16-1.3

8. The distance between two slits in a double-slit experiment is 0.005 mm. What is the angle of the third-order bright fringe ($m = 3$) produced with light of 550 nm?

   a. 5.0°
   b. 9.9°
   c. 12°
   d. 19°

   ANS: D   DIF: IIIB   OBJ: 16-1.3

9. The distance between two slits in a double-slit experiment is 2.9×10^{-6} m. The first-order bright fringe is measured on a screen at an angle of 12° from the central maximum. What is the wavelength of the light?

   a. 3.1×10^3 nm
   b. 1.2×10^2 nm
   c. 6.0×10^2 nm
   d. 4.6×10^2 nm

   ANS: C   DIF: IIIB   OBJ: 16-1.3

10. At the first dark band in a single-slit diffraction pattern, the path lengths of selected pairs of wavelets differ by

    a. one wavelength.
    b. more than one wavelength.
    c. one-half wavelength.
    d. less than half of one wavelength.

    ANS: C   DIF: IIIB   OBJ: 16-2.1

11. Monochromatic light shines on the surface of a diffraction grating with 5.0×10^3 lines/cm. The first-order maximum is observed at a 20.0° angle. Find the wavelength.

    a. 520 nm
    b. 480 nm
    c. 680 nm
    d. 360 nm

    ANS: C   DIF: IIIB   OBJ: 16-2.2

12. Light with a wavelength of 400.0 nm passes through a 1.00×10^4 lines/cm diffraction grating. What is the second-order angle of diffraction?

    a. 21.3°
    b. 53.1°
    c. 56.5°
    d. 72.1°

    ANS: B   DIF: IIIB   OBJ: 16-2.2

13. Light with a wavelength of 500.0 nm passes through a 3.39×10^5 lines/m diffraction grating. What is the first-order angle of diffraction?

    a. 23.5°
    b. 9.75°
    c. 36.9°
    d. 53.1°

    ANS: B   DIF: IIIB   OBJ: 16-2.2

14. Light with a wavelength of 546.1 nm passes through a 6.62×10^3 lines/cm diffraction grating. What is the first-order angle of diffraction?
   a. 21.2°   c. 39.2°
   b. 34.6°   d. 41.6°

   ANS: A   DIF: IIIB   OBJ: 16-2.2

15. The angle between the first-order maximum and the zeroth-order maximum for monochromatic light of 2300 nm is 27°. Calculate the number of lines per centimeter on this grating.
   a. 1600 lines/cm   c. 2500 lines/cm
   b. 2.0×10^3 lines/cm   d. 4500 lines/cm

   ANS: B   DIF: IIIB   OBJ: 16-2.2

16. For high resolution in optical instruments, the angle between resolved objects should be
   a. as small as possible.   c. 1.22°.
   b. as large as possible.   d. 45°.

   ANS: A   DIF: I   OBJ: 16-2.3

17. If light waves are coherent,
   a. they shift over time.
   b. their intensity is less than that of incoherent light.
   c. they remain in phase.
   d. they have less than three different wavelengths.

   ANS: C   DIF: I   OBJ: 16-3.1

18. Energy can be added to
   a. stimulated emission.   c. light amplification.
   b. an active medium.   d. a ray.

   ANS: B   DIF: I   OBJ: 16-3.1

19. Which of the following is the process of using a light wave to produce more waves with properties identical to those of the first wave?
   a. stimulated emission   c. hologram
   b. active medium   d. bandwidth

   ANS: A   DIF: I   OBJ: 16-3.1

20. Which of the following is a device that produces an intense, nearly parallel beam of coherent light?
   a. spectroscope   c. laser
   b. telescope.   d. diffraction grating

   ANS: C   DIF: I   OBJ: 16-3.1

21. A laser CANNOT be used
    a. to treat glaucoma.
    b. to measure distance.
    c. to read bar codes.
    d. to reverse heart-attack damage.

    ANS: D   DIF: I   OBJ: 16-3.2

SHORT ANSWER

1. The dark lines in a double-slit interference pattern are due to _____ interference, and the bright lines are due to _____ interference.

   ANS:
   destructive; constructive

   DIF: I   OBJ: 16-1.1

2. How is a stable interference pattern produced?

   ANS:
   Two sources are needed. Each source must produce a traveling periodic wave and the phases of the individual waves with respect to one another must be constant.

   DIF: I   OBJ: 16-2.2

3. Describe the pattern that results from the single-slit diffraction of monochromatic light.

   ANS:
   The pattern is one of alternation light and ark bands. The brightest light band is at the center. The light bands decrease in brightness as the distance from the center increases.

   DIF: IIIB   OBJ: 16-2.1

4. What is diffraction?

   ANS:
   Diffraction is the spreading of light into a region behind an obstruction.

   DIF: IIIB   OBJ: 16-2.1

5. How does diffraction occur?

   ANS:
   Diffraction occurs when waves pass through small openings, around obstacles, or around sharp edges. Wavelets in a wave front interfere with each other to produce light and dark fringes.

   DIF: I   OBJ: 16-1.3

6. What is resolving power?

ANS:
Resolving power is the ability of an optical instrument to separate two images that are close together.

DIF: I   OBJ: 16-2.3

7. Why is the resolving power for optical telescopes on Earth limited?

ANS:
Constantly moving layers of air blur the light from objects in space and limit the resolving power.

DIF: I   OBJ: 16-2.3

8. What is meant by the statement that a laser produces a narrow beam of coherent light?

ANS:
The waves emitted by a laser do not shift relative to each other as time progresses. The individual waves behave like a single wave because they are coherent and in phase.

DIF: II   OBJ: 16-3.1

9. How does a laser produce coherent light?

ANS:
When energy is added to the active medium, the atoms in the active medium absorb some of the energy. Later, these atoms release energy in the form of light waves that have the equivalent wavelength and phase. The initial waves cause other energized atoms to release their excess energy in the form of more light waves with the same wavelength, phase, and direction as the initial light wave. Mirrors on the end of the material return these coherent light waves into the active medium, where they emit more coherent light waves. One of these mirrors is slightly transparent so that some of the coherent light is emitted.

DIF: II   OBJ: 16-3.1

10. What are the advantages of using a "laser knife" in surgical procedures?

ANS:
A "laser knife" cuts through tissue like a steel scalpel; however, the energy from the laser coagulates blood, sealing the blood vessels and preventing blood loss and infection.

DIF: I   OBJ: 16-3.2

11. How is a laser used in a compact disc player?

     ANS:
     Light from a laser passes through a glass plate and then a lens that directs it onto the compact
     disc. The reflection or lack of a reflection of the laser light is then read by a detector and the
     signal is sent through electrical circuits.

     DIF: I   OBJ: 16-3.2

12. How are lasers used to determine the distance from Earth to the moon?

     ANS:
     Astronomers direct a pulse of light toward one of several reflectors that were placed on the
     moon's surface by astronauts. The distance from Earth to the moon can be measured by finding
     the time the light takes to travel to the moon and back.

     DIF: I   OBJ: 16-3.2

PROBLEM

1. Monochromatic light from a helium-neon laser ($\lambda = 632.8$ nm) shines at a right angle onto the
    surface of a diffraction grating that contains 130 960 lines/m. Find the angles at which one would
    observe the first-order and second-order maxima.

    ANS:
    $\theta_1 = 4.754°$; $\theta_2 = 9.540°$

    DIF: IIIB   OBJ: 16-2.2

2. Monochromatic light from a helium-neon laser ($\lambda = 632.8$ nm) shines at a right angle onto the
    surface of a diffraction grating that contains 149 638 lines/m. Find the angles at which one would
    observe the first-order and second-order maxima.

    ANS:
    $\theta_1 = 5.433°$; $\theta_2 = 10.92°$

    DIF: IIIB   OBJ: 16-2.2

3. Monochromatic light from a helium-neon laser ($\lambda = 632.8$ nm) shines at a right angle onto the
    surface of a diffraction grating that contains 630 692 lines/m. Find the angles at which one would
    observe the first-order and second-order maxima.

    ANS:
    $\theta_1 = 23.52°$; $\theta_2 = 10.66°$

    DIF: IIIB   OBJ: 16-2.2

4. Monochromatic light from a helium-neon laser ($\lambda = 632.8$ nm) shines at a right angle onto the surface of a diffraction grating that contains 146 159 lines/m. Find the angles at which one would observe the first-order and second-order maxima.

    ANS:
    $\theta_1 = 5.307°$; $\theta_2 = 10.66°$

    DIF: IIIB   OBJ: 16-2.2

5. Monochromatic light from a helium-neon laser ($\lambda = 632.8$ nm) shines at a right angle onto the surface of a diffraction grating that contains 787 412 lines/m. Find the angles at which one would observe the first-order and second-order maxima.

    ANS:
    $\theta_1 = 29.89°$; $\theta_2 = 85.24°$

    DIF: IIIB   OBJ: 16-2.2

6. Monochromatic light from a helium-neon laser ($\lambda = 632.8$ nm) shines at a right angle onto the surface of a diffraction grating that contains 395 013 lines/m. Find the angles at which one would observe the first-order and second-order maxima.

    ANS:
    $\theta_1 = 14.47°$; $\theta_2 = 29.99°$

    DIF: IIIB   OBJ: 16-2.2

7. Monochromatic light from a helium-neon laser ($\lambda = 632.8$ nm) shines at a right angle onto the surface of a diffraction grating that contains 462 138 lines/m. Find the angles at which one would observe the first-order and second-order maxima

    ANS:
    $\theta_1 = 17.00°$; $\theta_2 = 35.79°$

    DIF: IIIB   OBJ: 16-2.2

8. Monochromatic light from a helium-neon laser ($\lambda = 632.8$ nm) shines at a right angle onto the surface of a diffraction grating that contains 531 001 lines/m. Find the angles at which one would observe the first-order and second-order maxima.

    ANS:
    $\theta_1 = 19.63°$; $\theta_2 = 42.22°$

    DIF: IIIB   OBJ: 16-2.2

Holt Physics Assessment Item Listing
170

9. Monochromatic light from a helium-neon laser ($\lambda = 632.8$ nm) shines at a right angle onto the surface of a diffraction grating that contains 650 472 lines/m. Find the angles at which one would observe the first-order and second-order maxima

ANS:
$\theta_1 = 24.31°$; $\theta_2 = 55.41°$

DIF: IIIB   OBJ: 16-2.2

10. Monochromatic light from a helium-neon laser ($\lambda = 632,8$ nm) shines at a right angle onto the surface of a diffraction grating that contains 146 230 lines/m. Find the angles at which one would observe the first-order and second-order maxima.

ANS:
$\theta_1 = 5.309°$; $\theta_2 = 10.66°$

DIF: IIIB   OBJ: 16-2.2

MULTIPLE CHOICE

1. What happens when a rubber rod is rubbed with a piece of fur, giving it a negative charge?
   a. Protons are removed from the rod.
   b. Electrons are added to the rod.
   c. The fur is also negatively charged.
   d. The fur is left neutral.

   ANS: B   DIF: I   OBJ: 17-1.1

2. A repelling force occurs between two charged objects when
   a. charges are of unlike signs.
   b. charges are of like signs.
   c. charges are of equal magnitude.
   d. charges are of unequal magnitude.

   ANS: B   DIF: I   OBJ: 17-1.1

3. An attracting force occurs between two charged objects when
   a. charges are of unlike signs.
   b. charges are of like signs.
   c. charges are of equal magnitude.
   d. charges are of unequal magnitude.

   ANS: A   DIF: I   OBJ: 17-1.1

4. When a glass rod is rubbed with silk and becomes positively charged,
   a. electrons are removed from the rod.
   b. protons are removed from the silk.
   c. protons are added to the silk.
   d. the silk remains neutral.

   ANS: A   DIF: I   OBJ: 17-1.1

5. Electric charge is
   a. found only in a conductor.
   b. conserved.
   c. found only in insulators.
   d. not conserved.

   ANS: B   DIF: I   OBJ: 17-1.1

6. If a positively charged glass rod is used to charge a metal bar by induction,
   a. the charge on the bar will be equal in magnitude to the charge on the glass rod.
   b. the charge on the bar must be negative.
   c. the charge on the bar must be positive.
   d. the charge on the bar will be greater in magnitude than the charge on the glass rod.

   ANS: B   DIF: II   OBJ: 17-1.1

7. Which of the following transfers charge most easily?
   a. nonconductors
   b. conductors
   c. semiconductors
   d. insulators

   ANS: B   DIF: I   OBJ: 17-1.2

8. Which sentence best characterizes electric conductors?
   a. They have low mass density.
   b. They have high tensile strength
   c. They have electric charges that move freely.
   d. They are poor heat conductors.

   ANS: C   DIF: I   OBJ: 17-1.2

9. Which sentence best characterizes electric insulators?
   a. Charges on their surface do not move.   c. Electric charges move freely in them.
   b. They have high tensile strength   d. They are good heat conductors.

   ANS: A   DIF: I   OBJ: 17-1.2

10. The process of charging a conductor by bringing it near another charged object and then grounding the conductor is called
    a. charging by contact.   c. charging by polarization
    b. induction.   d. neutralization.

    ANS: B   DIF: I   OBJ: 17-1.3

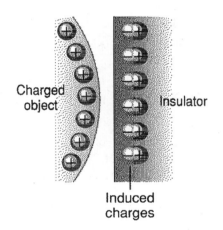

Charged object   Insulator

Induced charges

11. The figure above demonstrates charging by
    a. grounding.   c. polarization.
    b. induction.   d. contact.

    ANS: C   DIF: I   OBJ: 17-1.3

12. Both insulators and conductors can be charged by
    a. grounding.   c. polarization.
    b. induction   d. contact.

    ANS: D   DIF: I   OBJ: 17-1.3

13. A surface charge can be produced on insulators by
    a. grounding.   c. polarization.
    b. induction.   d. contact.

    ANS: D   DIF: I   OBJ: 17-1.3

13. A surface charge can be produced on insulators by
   a. grounding.
   b. induction.
   c. polarization.
   d. contact.

   ANS: D   DIF: I   OBJ: 17-1.3

14. Unlike insulators, conductors can be charged by
   a. grounding.
   b. induction.
   c. polarization.
   d. contact.

   ANS: B   DIF: I   OBJ: 17-1.3

15. When a charged body is brought close to an uncharged body without touching it, a(n) _____ charge may result on the uncharged body. When a charged body is brought into contact with an uncharged body and then is removed, a(n) _____ charge may result on the uncharged body.
   a. negative; positive
   b. positive; negative
   c. induced; residual
   d. residual; induced

   ANS: C   DIF: I   OBJ: 17-1.3

16. If two point charges are separated by 1.5 cm and have charge values of 2.0 μC and -4.0 μC, respectively, what is the value of the mutual force between them? ($k_c = 8.99 \times 10^9$ N•m^2/C^2)
   a. 320 N
   b. 3.6×10^{-8} N
   c. 8.0×10^{-12}
   d. 3.1×10^{-3} N

   ANS: A   DIF: IIIB   OBJ: 17-2.1

17. Consider a thundercloud that has an electric charge of 40.0 C near the top of the cloud and -40.0 C near the bottom of the cloud. These charges are separated by about 2.0 km. What is the electric force between these two sets of charges? ($k_c = 8.99 \times 10^9$ N•m^2/C^2)
   a. 3.6×10^4 N
   b. 3.6×10^5 N
   c. 3.6×10^6 N
   d. 3.6×10^7 N

   ANS: B   DIF: IIIB   OBJ: 17-2.1

18. An alpha particle (charge 2e) is sent at high speed toward a gold nucleus (charge 79e). What is the electric force acting on the alpha particle when it is 2.0×10^{-14} m away from the gold nucleus? ($e = 1.60 \times 10^{-9}$ C, $k_c = 8.99 \times 10^9$ N•m^2/C^2)
   a. 91 N
   b. 0.91 N
   c. 9.1×10^{-4} N
   d. 9.1×10^{-6} N

   ANS: A   DIF: IIIB   OBJ: 17-2.1

19. Which of the following is NOT true for BOTH gravitational and electric forces?
    a. The inverse square distance law applies.
    b. Forces are conservative.
    c. Potential energy is a function of distance of separation.
    d. Forces are either attractive or repulsive.

    ANS: D   DIF: I   OBJ: 17-2.2

20. Two point charges, initially 2 cm apart, are moved to a distance of 10 cm apart. By what factor do the resulting electric and gravitational forces between them change?
    a. 5   c. $\frac{1}{5}$

    b. 25   d. $\frac{1}{25}$

    ANS: D   DIF: I   OBJ: 17-2.2

21. If the charge and mass are tripled for two identical charges maintained at a constant separation, the electric and gravitational forces between them will be changed by what factor?
    a. 9   c. $\frac{1}{9}$

    b. $\frac{2}{3}$ d. 18

    ANS: A   DIF: I   OBJ: 17-2.2

22. Two point charges, initially 1 cm apart, are moved to a distance of 3 cm apart. By what factor do the resulting electric and gravitational forces between them change?
    a. 3   c. $\frac{1}{3}$

    b. 9   d. $\frac{1}{9}$

    ANS: D   DIF: I   OBJ: 17-2.2

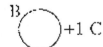

23. Four charges—A, B, C, and D— are at the corners of a square. Charges A and D, on opposite corners, have equal charge, whereas both B and C have a charge of 1.0 C. If the force on B is zero, what is the charge on A?
    a. −1.0 C   c. −0.35 C
    b. −0.20 C   d. −0.71 C

    ANS: C   DIF: IIIC   OBJ: 17-2.3

Holt Physics Assessment Item Listing
175

24. Charge A and charge B are 2 m apart. Charge A is 1 C and charge B is 2 C. Charge C, which is 2 C, is located between them, and the force on charge C is zero. How far from charge A is charge C?
   a. 1 m   c. 0.8 m
   b. 0.7 m   d. 0.5 m

   ANS: C   DIF: IIIC   OBJ: 17-2.3

25. Two charges are located on the positive x-axis of a coordinate system. Charge $q_1 = 2.00 \times 10^{-9}$ C, and it is 0.02 m from the origin. Charge $q_2 = -3.00 \times 10^{-9}$ C, and it is 0.04 m from the origin. What is the electric force exerted by these two charges on a third charge, $q_3 = 5.00 \times 10^{-9}$, located at the origin?
   a. 2.2×10^{-4} N   c. 3.1×10^{-4} N
   b. 1.4×10^{-4} N   d. 8.4×10^{-4} N

   ANS: B   DIF: IIIC   OBJ: 17-2.3

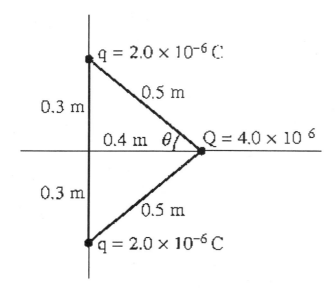

26. Two equal positive charges, both $q = 2.0 \times 10^{-6}$ C, interact with a third charge, $Q = 4.0 \times 10^{-6}$ C, as shown in the figure above. What is the magnitude of the electric force on Q?
   a. 0.23 N   c. 0.46 N
   b. −0.17 N   d. 0.60 N

   ANS: C   DIF: IIIC   OBJ: 17-2.3

27. Four point charges are positioned on the rim of a circle with a radius of 10 cm. The charges are 0.5 μC, 1.5 μC, −1.0 μC, and −0.5 μC, respectively. If the electric potential at the center of the circle due to the 0.5 charge alone is 4.5×10^4 V, what is the total electric force at the center due to the four charges combined? (Hint: Use the superposition principle.)
   a. 1.80×10^4 V   c. 0.0 V
   b. 4.5×10^4 V   d. -4.5×10^4 V

   ANS: A   DIF: IIIB   OBJ: 17-2.3

28. Two positive charges, each of magnitude q, are on the y-axis at points $y = +a$ and $y = -a$. Where would a third positive charge of the same magnitude be located for the net force on the third charge to be zero?
   a. at the origin
   b. at $y = 2a$
   c. at $y = -2a$
   d. at $y = -a$

   ANS: A   DIF: II   OBJ: 17-2.3

29. Two point charges have a value of 30 μC each and are 4 cm apart. What is the electric field at the midpoint between the two charges? ($k_c = 8.99 \times 10^9$ N•m^2/C^2)
   a. 4.5×10^7 N/C
   b. 2.3×10^7 N/C
   c. $.5 \times 10^7$ N/C
   d. 0 N/C

   ANS: D   DIF: IIIB   OBJ: 17-3.1

30. Two point charges are 10.0 cm apart and have charges of 2.0 μC and -2.0 μC, respectively. What is the electric field at the midpoint between the two charges?
   a. 2.9×10^7 N/C
   b. 1.4×10^7 N/C
   c. 7.2×10^6 N/C
   d. 0 N/C

   ANS: B   DIF: IIIB   OBJ: 17-3.1

31. Charges of 4.0 μC and -6.0 μC are placed at two corners of an equilateral triangle with sides of 0.10 m. At the third corner, what is the electric field magnitude created by these two charges?
   a. 4.8×10^6 N/C
   b. 3.1×10^6 N/C
   c. 1.6×10^6 N/C
   d. 7.4×10^6 N/C

   ANS: A   DIF: IIIC   OBJ: 17-3.1

32. The relative distribution of charge density on the surface of a conducting solid depends upon which of the following?
   a. the shape of the conductor
   b. the mass density of the conductor
   c. the type of metal the conductor is made of
   d. the strength of Earth's gravitational field

   ANS: A   DIF: I   OBJ: 17-3.3

33. At what point is the electric field of an isolated, uniformly charged, hollow metallic sphere greatest?
   a. at the center of the sphere
   b. at the sphere's inner surface
   c. at infinity
   d. at the sphere's outer surface

   ANS: D   DIF: I   OBJ: 17-3.3

34. If a conductor is in electrostatic equilibrium,
    a. the total charge on the conductor must be zero.
    b. the electric field inside the conductor must be zero.
    c. any charges on the conductor must be uniformly distributed.
    d. the conductor is grounded.

    ANS: B   DIF: IIIB   OBJ: 17-3.3

35. If a conductor is in electrostatic equilibrium, the electric field inside the conductor
    a. is directed inward.   c. is at its maximum level.
    b. is directed outward.   d. is zero.

    ANS: D   DIF: IIIB   OBJ: 17-3.3

36. If a conductor is in electrostatic equilibrium, the electric field just outside a charged conductor
    a. is zero.
    b. is at its minimum level.
    c. is the same as it is in the center of the conductor.
    d. is perpendicular to the conductor's surface.

    ANS: D   DIF: IIIB   OBJ: 17-3.3

37. If a conductor is in electrostatic equilibrium, any excess charge
    a. flows to the ground.
    b. resides entirely on the conductor's outer surface.
    c. resides entirely on the conductor's interior.
    d. resides entirely in the center of the conductor.

    ANS: B   DIF: IIIB   OBJ: 17-3.3

38. If an irregularly-shaped conductor is in electrostatic equilibrium, charge accumulates
    a. where the radius of curvature is smallest.   c. evenly throughout the conductor.
    b. where the radius of curvature is largest.   d. in flat places.

    ANS: A   DIF: IIIB   OBJ: 17-3.3

SHORT ANSWER

1. Explain what happens when you vigorously rub your wool socks on a carpeted floor, touch a metal doorknob, and get a shock.

    ANS:
    Loosely held electrons are transferred from the carpet to the socks when the socks are rubbed against the carpet. At that point, the body and socks have an excess of electrons and are negatively charged. Touching the doorknob allows the electrons to escape. The shock felt is the sudden movement of charges as the body and socks return to a neutral state.

    DIF: II   OBJ: 17:1.1

2. What property was discovered in Millikan's experiments? Explain this property.

ANS:
Millikan discovered that charge is quantized. This means that when any object is charged, the net charge is always a multiple of a fundamental unit of charge. The fundamental unit of charge, which is the charge on the electron, is -1.60×10^{-19} C. The charge on a proton is 1.60×10^{-19} C.

DIF:　I　　　　　OBJ:　17:1.1

3. When a conductor is given a negative charge, the charge will move on the conductor until the repulsive forces between the free electrons are in _____.

ANS:
equilibrium

DIF:　I　　　　　OBJ:　17:1.2

4. A large conducting sphere with a large net negative charge is brought close to, but does not touch, a small, uncharged conducting small sphere. A wire connects the sphere to a metal plate. If the wire to the metal plate is disconnected and the large sphere is removed, what are the charges on the smaller sphere and on the metal plate?

ANS:
positive on the sphere negative on the plate

DIF:　I　　　　　OBJ:　17:1.

5. What is electric force?

ANS:
Electric force is a field force exerted by one charge on another object. This force is attractive between opposite charges and repulsive between like charges.

DIF:　I　　　　　OBJ:　17-2.

6. How are gravitational and electric force alike?

ANS:
Both are field forces, and both have inverse square laws.

DIF:　I　　　　　OBJ:　17-2.

7. How are gravitational and electric force different?

ANS:
Electric force is stronger than gravitational force. Also, electric force can be attractive or repulsive, but gravitational force is always attractive.

DIF:　I　　　　　OBJ:　17-2.

8. Draw the lines of force representing the electric field around two equal but oppositely charged points that are 1.0 m apart in a vacuum. Each charge has a magnitude of 3.2×10^{-6} C. What is the intensity of the electric field at the midpoint between the charges? Assume that the resultant field is the sum of the fields caused by each charge. ($k_c = 8.99 \times 10^9$ N•m²/C²)

ANS:
2.3×10^5 N/C

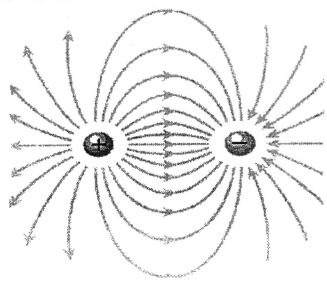

DIF: I   OBJ: 17-3.2

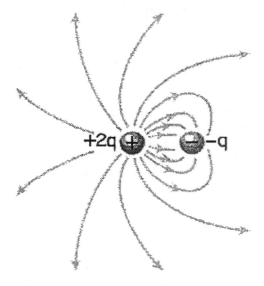

9. In the figure shown above, why do only half of the lines originating from the positive charge terminate on the negative charge?

ANS:
Because the positive charge is twice as great as the negative charge.

DIF: I   OBJ: 17-3.2

10.   Draw the lines of force representing the electric field around a proton.

ANS:

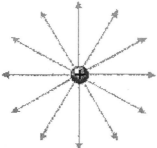

DIF: I   OBJ: 17-3.2

11. Draw the lines of force representing the electric field around an electron.

ANS:

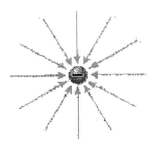

DIF: I   OBJ: 17-3.2

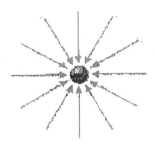

12. Is the charge shown in the figure above positive or negative?

ANS:
negative

DIF: I   OBJ: 17-3.2

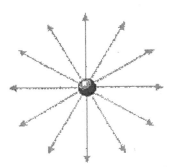

13. Is the charge shown in the figure above positive or negative?

ANS:
positive

DIF: I   OBJ: 17-3.2

Holt Physics Assessment Item Listing
182

14. Draw the lines of force representing the electric field around two equal but oppositely charged points in a vacuum. The charges each have a magnitude of 1.6×10^{-19} C, and they are 1×10^{-11} m apart. Assume that the resultant field is the sum of the fields caused by each charge.

ANS:

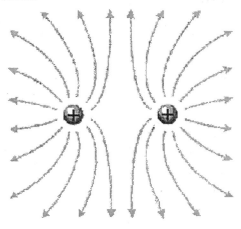

DIF: I   OBJ: 17-3.2

PROBLEM

1. What is the electric force between an electron and a proton that are separated by a distance of 1.0×10^{-10} m? ($e = 1.60 \times 10^{-19}$ C, $k_c = 8.99 \times 10^9$ N•m²/C²)

ANS:
2.3×10^{-8} N

DIF: IIIA   OBJ: 17-2.1

2. What is the electric force between a proton and an alpha particle (charge $2e$) that are separated by a distance of 3.0×10^{-6} m? ($e = 1.60 \times 10^{-19}$ C, $k_c = 8.99 \times 10^9$ N•m²/C²)

ANS:
5.1×10^{-17} N

DIF: IIIA   OBJ: 17-2.1

3. What is the electric force between an electron and a potassium nucleus (charge $19e$) that are separated by a distance of 5.2×10^{-10} m? ($e = 1.60 \times 10^{-19}$ C, $k_c = 8.99 \times 10^9$ N•m²/C²)

ANS:
1.6×10^{-8} N

DIF: IIIA   OBJ: 17-2.1

4. What is the electric force between an alpha particle (charge $2e$) and an oxygen nucleus (charge $8e$) that are separated by a distance of 3.6×10^{-6} m? ($e = 1.60 \times 10^{-19}$ C, $k_c = 8.99 \times 10^9$ N•m²/C²)

ANS:
2.8×10^{-6} N

DIF: IIIA   OBJ: 17-2.1

5. What is the electric force between an electron and an alpha particle (charge $2e$) that are separated by a distance of 3.5×10^{-3} m? ($e = 1.60 \times 10^{-19}$ C, $k_c = 8.99 \times 10^9$ N•m²/C²)

ANS:
3.8×10^{-23} N

DIF: IIIA   OBJ: 17-2.1

6. What is the electric force between an alpha particle (charge $2e$) and an carbon nucleus (charge $6e$) that are separated by a distance of 7.8×10^{-4} m? ($e = 1.60 \times 10^{-19}$ C, $k_c = 8.99 \times 10^9$ N•m²/C²)

ANS:
4.5×10^{-21} N

DIF: IIIA   OBJ: 17-2.1

7. What is the electric force between two electrons that are separated by a distance of 5.9×10^{-8} m? ($e = 1.60 \times 10^{-19}$ C, $k_c = 8.99 \times 10^9$ N•m²/C²)

ANS:
6.6×10^{-14} N

DIF: IIIA   OBJ: 17-2.1

MULTIPLE CHOICE

1. Which of the following is NOT a characteristic of electrical potential energy?
   a. It is a form of mechanical energy.
   b. It results from a single charge.
   c. It results from the interaction between charges.
   d. It is associated with a charge in an electric field.

   ANS: B   DIF: I   OBJ: 18-1.1

2. When a positive charge moves because of a force, what happens to the electrical potential energy associated with the charge's position in the system?
   a. It increases.   c. It remains the same.
   b. It decreases.   d. It sharply increases, and then decreases.

   ANS: B   DIF: I   OBJ: 18-1.1

3. Two positive point charges are initially separated by a distance of 2 cm. If their separation is increased to 6 cm, the resultant electrical potential energy is equal to what factor times the initial electrical potential energy?
   a. 3   c. $\frac{1}{3}$

   b. 9   d. $\frac{1}{9}$

   ANS: C   DIF: II   OBJ: 18-1.2

4. A proton ($q = 1.60 \times 10^{19}$ C) moves 10.0 cm on a path parallel to the direction of a uniform electric field of strength 3.0 N/C. What is the change in electrical potential energy?
   a. 4.8×10^{-20} J   c. -4.8×10^{-20} J
   b. 1.6×10^{-20} J   d. -1.6×10^{-20} J

   ANS: C   DIF: IIIB   OBJ: 18-1.2

5. A uniform electric field with a magnitude of 5.0×10^2 N/C is directed parallel to the positive x-axis toward the origin. What is the change in electrical energy of a proton ($q = 1.60 \times 10^{-19}$ C) as it moves from $x = 5$ m to $x = 2$ m?
   a. 8.0×10^{-17} J   c. 2.0×10^{21} J
   b. 2.0×10^{-16} J   d. 500 J

   ANS: B   DIF: IIIB   OBJ: 18-1.2

6. Two point charges with values of 3.4 μC and 6.6 C are separated by 0.20 m. What is the electrical potential energy of this two-charge system?
   a. 0.34 J   c. 1.0 J
   b. −0.75 J   d. −3.4 J

   ANS: C   DIF: IIIB   OBJ: 18-1.2

7. Two protons, each having a charge of 1.60×10^{-19} C, are 2.0×10^{-5} m apart. What is the electrical potential energy between the two charges?
    a. 1.1×10^{-23} J
    b. 3.2×10^{-19} J
    c. 3.2×10^{-16} J
    d. 1.6×10^{-14} J

    ANS: A   DIF: IIIB   OBJ: 18-1.2

8. When an electron ($e = -1.6 \times 10^{-19}$ C) moves 0.10 m along the direction of an electric field with a strength of 3.0 N/C, what is the magnitude of the potential difference between the electron's initial and final points?
    a. 4.8×10^{-19} V
    b. 0.30 V
    c. 0.03 V
    d. 3.0×10^1 V

    ANS: B   DIF: IIIB   OBJ: 18-2.2

9. Four point charges are positioned on the circumference of a circle with a radius of 10 cm. The charges are 0.5 μC, 1.5 μC, –1.0 μC, and –0.5 μC, respectively. If the electric potential at the center of the circle due to the 0.5 charge alone is 4.5×10^4 V, what is the total potential at the center due to the four charges combined? (Hint: Use the superposition principle.)
    a. 1.80×10^4 V
    b. 4.5×10^4V
    c. 0.0 V
    d. -4.5×10^4 V

    ANS: B   DIF: IIIB   OBJ: 18-2.2

10. A uniform electric field with a magnitude of 500 N/C is directed parallel to the positive x-axis. If the potential at $x = 5$ m is 2500 V, what is the potential at $x = 2$ m?
    a. 1000 V
    b. 2000 V
    c. 4000 V
    d. 4500 V

    ANS: C   DIF: IIIB   OBJ: 18-2.2

11. What will be the electric potential at a distance of 0.15 m from a point charge of 6.0 μC?
    ($k_c = 8.99 \times 10^9$ N•m^2/C^2)
    a. 5.4×10^4 V
    b. 3.6×10^6 V
    c. 2.4×10^6 V
    d. 1.2×10^7 V

    ANS: B   DIF: IIIB   OBJ: 18-2.2

12. Two point charges with values of 3.4 μC and 6.6 μC are separated by 0.10 m. What is the electric potential at the point midway between the two point charges? ($k_c = 8.99 \times 10^9$ N•m^2/C^2)
    a. 1.8×10^6 V
    b. -0.9×10^6 V
    c. 0.9×10^6 V
    d. 3.6×10^6 V

    ANS: B   DIF: IIIB   OBJ: 18-2.2

13. At what distance from a point charge of 8.0 μC would the electric potential be 4.2×10^4 V?
    ($k_c = 8.99 \times 10^9$ N•m^2/C^2)
    a. 0.58 m
    b. 0.76 m
    c. 1.7 m
    d. 2.9 m

    ANS: C   DIF: IIIB   OBJ: 18-2.2

14. A point charge of 3.0 μC is at the origin of a coordinate system, and a second point charge of –6.0 μC is at $x = 1.0$ m. What is the electric potential at the point where $x = 0.50$ m? ($k_c = 8.99 \times 10^9$ N•m^2/C^2)
    a. 1.62×10^5 V   c. -1.08×10^5 V
    b. 1.08×10^5 V   d. -5.4×10^4 V

    ANS: D   DIF: IIIB   OBJ: 18-2.2

15. Charge build up between the plates of a capacitor stops when
    a.   there is no net charge on the plates.
    b.   unequal amounts of charge accumulate on the plate.
    c.   the potential difference between the plates is equal to the potential difference between the terminals of the battery.
    d.   the charge on both plates is the same.

    ANS: C   DIF: II   OBJ: 18-3.1

16. When comparing the net charge of a charged capacitor with the net charge of the same capacitor when it is uncharged, the net charge is
    a.   greater in the charged capacitor.
    b.   less in the charged capacitor.
    c.   equal in both capacitors.
    d.   greater or less in the charged capacitor, but never equal.

    ANS: C   DIF: I   OBJ: 18-3.1

17. When a capacitor discharges,
    a.   it must be attached to a battery.
    b.   charges move back from one plate to another through the circuit until both plates are uncharged.
    c.   charges move from one plate to another until equal and opposite charges accumulate on the plates.
    d.   it cannot be connected to a material that conducts.

    ANS: B   DIF: I   OBJ: 18-3.1

18. A capacitor consists of two metal plates; _____ is stored on one plate and _____is stored on the other.
    a.   negative charge; positive charge   c.   potential difference; internal resistance
    b.   potential energy; kinetic energy   d.   residual charge; induced charge

    ANS: A   DIF: I   OBJ: 18-3.1

19. Increasing the separation of the two charged parallel plates of a capacitor when they are disconnected from a battery will produce what effect on the capacitor?
    a.   It will increase the charge.   c.   It will increase the capacitance.
    b.   It will decrease the charge.   d.   It will decrease the capacitance.

    ANS: D   DIF: I   OBJ: 18-3.1

20. Increasing the potential difference across the two plates of a capacitor will produce what effect on the capacitor?
    a. It will increase the charge.
    b. It will decrease the charge.
    c. It will increase the capacitance.
    d. It will decrease the capacitance.

    ANS: A   DIF: I   OBJ: 18-3.1

21. A 0.25 μF capacitor is connected to a 9.0 V battery. What is the charge on the capacitor?
    a. 1.2×10^{-12} C
    b. 2.2×10^{-6} C
    c. 2.8×10^{-8} C
    d. 3.6×10^{-7} C

    ANS: B   DIF: I   OBJ: 18-3.2

22. A parallel-plate capacitor has a capacitance of C F. If the area of the plates is doubled while the distance between the plates is halved, the new capacitance will be
    a. 2 C.
    b. 4 C.
    c. $\dfrac{C}{2}$.
    d. $\dfrac{C}{4}$.

    ANS: B   DIF: I   OBJ: 18-3.2

23. What is the capacitance of a parallel-plate capacitor made of two square aluminum plates that are 4.0 cm in length on each side and are separated by 5.0 mm? ($\varepsilon_0 = 8.85 \times 10^{-12}$ C²/N•m²)
    a. 2.8×10^{-10} F
    b. 1.0×10^{-6} F
    c. 2.8×10^{-12} F
    d. 2.0×10^{-5} F

    ANS: C   DIF: IIIB   OBJ: 18-3.2

24. A 0.50 μF capacitor is connected to a 12 V battery. How much electrical potential energy is stored in the capacitor?
    a. 2.0×10^{-12} J
    b. 1.0×10^{-12} J
    c. 0.04 J
    d. 3.6×10^{-5} J

    ANS: D   DIF: IIIB   OBJ: 18-3.3

25. A 1.5 μF capacitor is connected to a 9.0 V battery. How much energy is stored in the capacitor?
    a. 1.7×10^{-3} J
    b. 6.1×10^{-5} J
    c. 7.5×10^{-3} J
    d. 5.4×10^{-3} j

    ANS: B   DIF: IIIB   OBJ: 18-3.3

SHORT ANSWER

1. What is electrical potential energy?

   ANS:
   Potential energy associated with a charge due to the position of the charge in an electric field

   DIF: I   OBJ: 18-1.1

2. Electrical potential energy is a result of what interaction?

   ANS:
   The interaction between two charged objects

   DIF: I   OBJ: 18-1.1

3. What is electric potential?

   ANS:
   It is the result of the electrical potential energy associated with a charged particle divided by the charge of the particle.

   DIF: I   OBJ: 18-2.1

4. What is potential difference?

   ANS:
   It is the change in the electrical potential energy associated with a charged particle divided by the charge of the particle.

   DIF: I   OBJ: 18-2.1

5. How are electric potential and electrical potential energy related?

   ANS:
   Electric potential is the electrical potential energy associated with a charged particle in an electric field divided by the charge of the particle.

   DIF: I   OBJ: 18-2.1

6. What is capacitance?

   ANS:
   Capacitance is the measure of the ability of a device to store charge for a given potential difference.

   DIF: I   OBJ: 18-3.1

7. How does a capacitor store energy?

ANS:
Charges on a capacitor plate repel any additional charges that are brought up to the plate. The work done on charges to place them on the plates is the energy that is stored in the capacitor.

DIF: I   OBJ: 18-3.1

8. Explain why there is a limit to the amount of charge that can be stored on a capacitor.

ANS:
The potential can become so great that a spark discharge or an electrical breakdown will occur.

DIF: II   OBJ: 18-3.1

9. List two ways to increase the electrical potential energy that can be stored in a capacitor.

ANS:
1. Increase the area of the plates. 2. Place a dielectric material between the plates.

DIF: I   OBJ: 18-3.3

PROBLEM

1. A proton ($q = 1.60 \times 10^{-19}$ C) moves 16.0 cm on a path parallel to the direction of a uniform electric field of strength 3.0 N/C. What is the change in electrical potential energy?

ANS:
-5.1×10^{-20} J

DIF: IIIB   OBJ: 18-1.2

2. A proton ($q = 1.60 \times 10^{-19}$ C) moves 26 cm on a path parallel to the direction of a uniform electric field of strength 4.0 N/C. What is the change in electrical potential energy?

ANS:
-1.7×10^{-19} J

DIF: IIIB   OBJ: 18-1.2

3. A proton ($q = 1.60 \times 10^{-19}$ C) moves 38 cm on a path parallel to the direction of a uniform electric field of strength 1.5 N/C. What is the change in electrical potential energy?

ANS:
-9.2×10^{-20} J

DIF: IIIB   OBJ: 18-1.2

4. A proton ($q = 1.60 \times 10^{-19}$ C) moves 12 cm on a path parallel to the direction of a uniform electric field of strength 6.0 N/C. What is the change in electrical potential energy?

ANS:
-1.1×10^{-19} J

DIF: IIIB   OBJ: 18-1.2

5. A proton ($q = 1.60 \times 10^{-19}$ C) moves 24 cm on a path parallel to the direction of a uniform electric field of strength 5.0 N/C. What is the change in electrical potential energy?

ANS:
-1.9×10^{-19} J

DIF: IIIB   OBJ: 18-1.2

6. A proton ($q = 1.60 \times 10^{-19}$ C) moves 96 cm on a path parallel to the direction of a uniform electric field of strength 1.0 N/C. What is the change in electrical potential energy?

ANS:
-1.5×10^{-19} J

DIF: IIIB   OBJ: 18-1.2

7. A proton ($q = 1.60 \times 10^{-19}$ C) moves 47 cm on a path parallel to the direction of a uniform electric field of strength 3.6 N/C. What is the change in electrical potential energy?

ANS:
-2.7×10^{-19} J

DIF: IIIB   OBJ: 18-1.2

8. A proton ($q = 1.60 \times 10^{-19}$ C) moves 73 cm on a path parallel to the direction of a uniform electric field of strength 2.2 N/C. What is the change in electrical potential energy?

ANS:
-2.6×10^{-19} J

DIF: IIIB   OBJ: 18-1.2

9. A proton ($q = 1.60 \times 10^{-19}$ C) moves 68 cm on a path parallel to the direction of a uniform electric field of strength 2.8 N/C. What is the change in electrical potential energy?

ANS:
-3.0×10^{-19} J

DIF: IIIB   OBJ: 18-1.2

Holt Physics Assessment Item Listing

10. A 3.2 μF capacitor has a potential difference of 21.0 V between its plates. How much additional charge flows into the capacitor if the potential difference is increased to 47.0 V?

ANS:
8.3×10^5 C

DIF: IIIA   OBJ: 18-3.2

11. A 0.63 μF capacitor is connected to a 3.0 V battery. How much energy is stored in the capacitor?

ANS:
12.8×10^{-6} J

DIF: IIIB   OBJ: 18-3.3

12. A 3.2 μF capacitor is connected to a 1.5 V battery. How much energy is stored in the capacitor?

ANS:
3.6×10^{-6} J

DIF: IIIB   OBJ: 18-3.3

13. A 6.0 μF capacitor holds 3.0 μC of charge. How much potential energy is stored in the capacitor?

ANS:
7.5×10^{-7} J

DIF: IIIB   OBJ: 18-3.3

14. A 0.75 μF capacitor holds 6.0 μC of charge. How much potential energy is stored in the capacitor?

ANS:
2.4×10^{-5} J

DIF: IIIB   OBJ: 18-3.3

15. A 0.10 μF capacitor holds 9.0 μC of charge. How much potential energy is stored in the capacitor?

ANS:
4.0×10^{-4} J

DIF: IIIB   OBJ: 18-3.3

Holt Physics Assessment Item Listing
192

16. A 0.42 μF capacitor holds 1.0 μC of charge. How much potential energy is stored in the capacitor?

ANS:
1.2×10^{-6} J

DIF: IIIB   OBJ: 18-3.3

17. A 0.5 F capacitor is connected to a 1.5 V battery. How much energy is stored in the capacitor?

ANS:
0.56 J

DIF: IIIB   OBJ: 18-3.3

18. A 0.10 F capacitor is connected to a 4.0 V battery. How much energy is stored in the capacitor?

ANS:
0.80 J

DIF: IIIB   OBJ: 18-3.3

MULTIPLE CHOICE

1. How is current affected if the number of charge carriers decreases?
    a. The current increases.
    b. The current decreases.
    c. The current initially decreases and then is gradually restored.
    d. The current is not affected.

    ANS: B   DIF: II   OBJ: 19-1.1

2. How is current affected if the time interval decreases while the amount of charge remains the same?
    a. The current increases.
    b. The current decreases.
    c. The current initially increases and then is gradually restored.
    d. The current is not affected.

    ANS: A   DIF: II   OBJ: 19-1.1

3. The current in an electron beam in a cathode-ray tube is 7.0×10^{-5} A. How much charge hits the screen in 5.0 s?
    a. 2.8×10^3 C   c. 3.5×10^{-4} C
    b. 5.6×10^{-23} C   d. 5.3×10^{-4} C

    ANS: C   DIF: IIIA   OBJ: 19-1.2

4. A wire carries a steady current of 0.1 A over a period of 20 s. What total charge moves through the wire in this time interval?
    a. 200 C   c. 2 C
    b. 20 C   d. 0.005 C

    ANS: C   DIF: IIIA   OBJ: 19-1.2

5. The amount of charge that moves through the filament of a light bulb in 2.00 s is 2.67 C. What is the current in the light bulb?
    a. 5.34 A   c. 0.835 A
    b. 1.33 A   d. 0.417 A

    ANS: B   DIF: IIIB   OBJ: 19-1.2

6. When you flip a switch to turn on a light, the delay time before the light turns on is determined by
    a. the number of electron collisions per second in the wire.
    b. the drift speed of the electrons in the wire.
    c. the speed of the electric field moving in the wire.
    d. the resistance of the wire.

    ANS: C   DIF: IIIB   OBJ: 19-1.3

7. When electrons move through a metal conductor,
   a. they move in a straight line through the conductor.
   b. they move in zigzag patterns because of repeated collisions with the vibrating metal atoms.
   c. the temperature of the conductor decreases.
   d. they move at the speed of light in a vacuum.

   ANS: B   DIF: I   OBJ: 19-1.3

8. The energy gained by electrons as they are accelerated by an electric field is
   a. greater than the average loss in energy due to collisions.
   b. equal to the average loss in energy due to collisions.
   c. less than the average loss in energy due to collisions.
   d. not affected by the gain in energy due to collisions.

   ANS: A   DIF: I   OBJ: 19-1.3

9. In a conductor that carries a current, the drift speed of an electron is
   a. less than the average speed of the electron between collisions.
   b. equal to the average speed of the electron between collisions.
   c. greater than the average speed of the electron between collisions.
   d. approximately equal to the speed of light.

   ANS: A   DIF: I   OBJ: 19-1.3

10. The drift velocity in a wire is
    a. the average speed of electrons between collisions.
    b. the energy gained by electrons as they are accelerated by an electric field.
    c. the speed at which an electric field reaches electrons throughout a conductor.
    d. the net velocity of a charge carrier moving in an electric field.

    ANS: D   DIF: I   OBJ: 19-1.3

11. In alternating current, the motion of the charges
    a. continuously changes in the forward and reverse directions.
    b. is equal to the speed of light.
    c. is greater than the speed of light.
    d. in the direction of the electric field.

    ANS: A   DIF: II   OBJ: 19-1.4

12. What is the potential difference across a resistor of 5.0 Ω that carries a current of 5.0 A?
    a. 1.0×10^2 V   c. 4.0 V
    b. 25 V   d. 1.0 V

    ANS: B   DIF: IIIA   OBJ: 19-2.1

13. A flashlight bulb with a potential difference of 4.5 V across it has a resistance of 8.0 Ω. How much current is in the bulb filament?
   a.   3.7 A   c.   9.4 A
   b.   1.8 A   d.   0.56 A

   ANS: D   DIF: IIIB   OBJ: 19-2.1

14. A light bulb has a resistance of 240 Ω when operating at 120 V. What is the current in the light bulb?
   a.   2.0 A   c.   0.50 A
   b.   1.0 A   d.   0.20 A

   ANS: C   DIF: IIIA   OBJ: 19-2.1

15. Which of the following does NOT affect a materialís resistance?
   a.   length   c.   temperature
   b.   the type of material   d.   Ohm's law

   ANS: D   DIF: I   OBJ: 19-2.3

16. Which of the following wires would have the LEAST resistance? (Assume that the wires have the same cross-sectional area.)
   a.   an aluminum wire 10 cm in length   c.   a copper wire 10 cm in length
   b.   an aluminum wire 5 cm in length   d.   a copper wire 5 cm in length

   ANS: D   DIF: I   OBJ: 19-2.3

17. Which of the following wires would have the GREATEST resistance?
   a.   an aluminum wire 10 cm in length and 3 cm in diameter
   b.   an aluminum wire 5 cm in length and 3 cm in diameter
   c.   an aluminum wire 10 cm in length and 5 cm in diameter
   d.   an aluminum wire 5 cm in length and 5 cm in diameter

   ANS: A   DIF: I   OBJ: 19-2.3

18. Which of the following wires would have the LEAST resistance?
   a.   a copper wire 10 cm in length at 32°C   c.   a copper wire 10 cm in length at 10°C
   b.   a copper wire 5 cm in length at 32°C   d.   a copper wire 5 cm in length at 10°C

   ANS: D   DIF: I   OBJ: 19-2.3

19. Which of the following wires would have the LEAST resistance?
   a.   an aluminum wire 20 cm in diameter at 40°C
   b.   an aluminum wire 20 cm in diameter at 60°C
   c.   an aluminum wire 40 cm in diameter at 40°C
   d.   an aluminum wire 40 cm in diameter at 60°C

   ANS: A   DIF: I   OBJ: 19-2.3

20. What happens to the resistance of a superconductor when its temperature drops below the critical temperature?
    a. The resistance is equal to that of a semiconductor with the same dimensions.
    b. The resistance doubles.
    c. The resistance drops to zero.
    d. The resistance reduces by one-half.

    ANS: C   DIF: I   OBJ: 19-2.4

21. Consider a material that is cooled until it becomes a superconductor. If it is cooled even further, its resistance will
    a. increase.   c. stay constant and nonzero.
    b. decrease.   d. remain at zero.

    ANS: D   DIF: I   OBJ: 19-2.4

22. When current is in a superconductor,
    a. the current continues even if the applied potential difference is removed.
    b. the current continues until the applied potential difference is removed.
    c. the current quickly decays unless it is powered by an external energy source.
    d. the superconductor MUST be above the critical temperature.

    ANS: A   DIF: I   OBJ: 19-2.4

23. Which of the following statements about superconductors is correct?
    a. Superconductors require a magnet to carry current.
    b. Steady currents have been observed to persist for many years with no apparent decay in superconducting loops.
    c. Copper, silver, and gold are excellent superconductors.
    d. The resistance-temperature graph for a superconductor resembles that of a normal metal at temperatures below the critical temperature.

    ANS: B   DIF: I   OBJ: 19-2.4

24. The power ratings on light bulbs are measures of the
    a. rate that they give off heat and light.
    b. voltage they require.
    c. density of the charge carriers.
    d. amount of negative charge passing through them.

    ANS: A   DIF: II   OBJ: 19-3.1

25. When compared in a given time interval with other light bulbs in a 120 V outlet, a 60 W light bulb
    a. converts the same electrical energy to heat and light than a 40 W light bulb.
    b. converts more electrical energy to heat and light than a 100 W light bulb.
    c. converts less electrical energy to heat and light than a 40 W light bulb.
    d. converts less electrical energy to heat and light than a 100 W light bulb.

    ANS: D   DIF: I

26. If a 75 W light bulb operates at a voltage of 120 V, what is the current in the bulb?
    a.   0.63 A
    b.   1.6 A
    c. 9.0×10^3 A
    d. 1.1×10^{-4} A

    ANS: A   DIF: IIIA   OBJ: 19-3.2

27. Tripling the current in a circuit with constant resistance has the effect of changing the power by what factor?
    a. $\frac{1}{3}$ c.   3

    b. $\frac{1}{9}$ d.   9

    ANS: D   DIF: II   OBJ: 19-3.2

28. If a 5.00×10^2 W heater has a current of 4.00 A, what is the potential difference across the ends of the heating element?
    a. 2.00×10^3 V   c. 2.50×10^2 V
    b.   125 V   d. 8.00×10^{-3} V

    ANS: B   DIF: IIIB   OBJ: 19-3.2

29. If a 325 W heater has a current of 6.0 A, what is the resistance of the heating element?
    a.   88 Ω c.   9.0 Ω
    b.   54 Ω d.   11 Ω

    ANS: C   DIF: IIIB   OBJ: 19-3.2

30. If a lamp has a resistance of 120 Ω when it operates at a power of 1.00×10^2 W, what is the potential difference across the lamp?
    a.   110 V   c.   130 V
    b.   120 V   d.   220 V

    ANS: A   DIF: IIIB   OBJ: 19-3.2

31. If a lamp is measured to have a resistance of 45 Ω when it operates at a power of 80 W, what is the current in the lamp?
    a.   2.10 A   c.   0.91 A
    b.   1.3 A   d.   0.83 A

    ANS: B   DIF: IIIB   OBJ: 19-3.2

32. An electric toaster requires 1100 W at 110 V. What is the resistance of the heating coil?
    a.   7.5 Ω c. 1.0×10^1 Ω
    b.   9.0 Ω d.   11 Ω

    ANS: D   DIF: IIIB   OBJ: 19-3.2

33. A steam turbine at an electric power plant delivers 4500 kW of power to an electrical generator that converts 95 percent of this mechanical energy into electrical energy. What is the current delivered by the generator if it delivers energy at 3600 V?
    a. 0.66×10^3 A
    b. 1.0×10^3 A
    c. 1.2×10^3 A
    d. 5.9×10^3 A

    ANS: C   DIF: IIIB   OBJ: 19-3.2

34. Which process will double the power dissipated by a resistor?
    a.   doubling the current while doubling the resistance
    b.   doubling the current and making the resistance half as big
    c.   doubling the current and doubling the potential difference
    d.   doubling the current while making the potential difference half as big

    ANS: B   DIF: IIIB   OBJ: 19-3.2

35. A high-voltage transmission line carries 1000 A at 700 000 V. What is the maximum power carried in the line?
    a.   700 MW
    b.   400 MW
    c.   100 MW
    d.   70 MW

    ANS: A   DIF: IIIB   OBJ: 19-3.2

36. How much does it cost to operate a 695 W heater for exactly 30.0 mm if electrical energy costs $0.060 per kW•h?
    a. $0.02
    b. $0.90
    c. $0.18
    d. $0.36

    ANS: A   DIF: IIIB   OBJ: 19-3.3

37. A color television draws about 2.5 A when it is connected to a 120 V outlet. Assuming electrical energy costs $0.06 per kW•h, what is the cost of running the TV for exactly 8 h?
    a. $.014
    b. $0.03
    c. $0.14
    d. $ 0.30

    ANS: C   DIF: IIIB   OBJ: 19-3.3

38. A microwave draws 5.0 A when it is connected to a 120 V outlet. If electrical energy costs $0.090/kW•h, what is the cost of running the microwave for exactly 6 h?
    a. $2.70
    b. $1.60
    c. $0.72
    d. $0.32

    ANS: D   DIF: IIIB   OBJ: 19-3.3

39. A hair dryer draws 11 A when it is connected to 120 V. If electrical energy costs $ 0.09/kW•h, what is the cost of using the hair dryer for exactly 15 min?
    a. $0.33
    b. $0.12
    c. $0.06
    d. $0.03

    ANS: D   DIF: IIIB   OBJ: 19-3.3

SHORT ANSWER

1. What is electric current?

    ANS:
    Current is the rate at which electric charges move through an area.

    DIF: I   OBJ: 19-1.1

2. What are some applications of electric current?

    ANS:
    Answers will vary. Electric currents power lights and home appliances. They also ignite the gasoline in automobile engines, and they are used in microcomputers.

    DIF: I   OBJ: 19-1.1

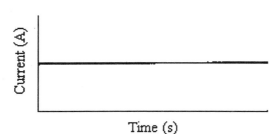

3. What type of electric current is shown in the figure above?

    ANS:
    Direct current

    DIF: I   OBJ: 19-1.4

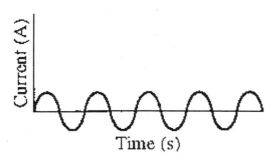

4. What type of electric current is shown in the figure above?

    ANS:
    Alternating current

    DIF: I   OBJ: 19-1.4

5. What are the characteristics of direct current?

    ANS:
    Charges move in only one direction.

    DIF: I   OBJ: 19-1.4

6. What are the characteristics of alternating current?

    ANS:
    The motion of charges continually changes in the forward and reverse directions. There is no net motion of the charge carriers; they simply vibrate back and forth.

    DIF: I   OBJ: 19-1.4

7. Which type of electric current is supplied to homes and businesses? Why?

    ANS:
    Alternating current is supplied because it is more practical for use in transferring electrical energy.

    DIF: I   OBJ: 19-1.4

8. What is an ohmic material?

    ANS:
    Any material that has a constant resistance over a wide range of potential differences is considered ohmic.

    DIF: I   OBJ: 19-2.2

9. What is a nonohmic material?

    ANS:
    A nonohmic material does not have constant resistance over a wide range of potential differences.

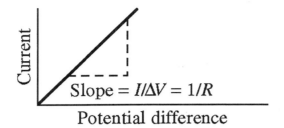

Slope = $I/\Delta V$ = $1/R$

Potential difference

10. What type of material does the current-potential difference curve illustrate in the graph above?

ANS:
ohmic material

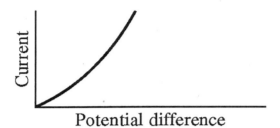

Potential difference

11. What type of material does the current-potential difference curve illustrate in the graph above?

ANS:
nonohmic material

12. What is electric power?

ANS:
Electric power is the rate at which charge carriers convert electric potential energy to nonelectrical forms of energy.

PROBLEM

1. A bolt of lightning discharges 1.7 C in 3.3×10^{-4} s. What is the average current during the discharge?

   ANS:
   15.1×10^3 A

   DIF: IIIA   OBJ: 19-1.2

2. The amount of charge that moves through the filament of a microwave in 10.0 s is 24.2 C. What is the current in the microwave?

   ANS:
   2.42 A

   DIF: IIIB   OBJ: 19-1.2

3. The amount of charge that moves through a blender in 5.0 s is 8.15 C. What is the current in the blender?

   ANS:
   1.63 A

   DIF: IIIB   OBJ: 19-1.2

4. The amount of charge that moves through the filament of a toaster in 45.0 s is 43.2 C. What is the current in the toaster?

   ANS:
   0.96 A

   DIF: IIIB   OBJ: 19-1.2

5. The amount of charge that moves through an electric fan in 15.0 s is 18.75 C. What is the current in the fan?

   ANS:
   1.25 A

   DIF: IIB   OBJ: 19-1.2

6. The amount of charge that moves through an electric juicer in 36.0 s is 48.96 C. What is the current in the juicer?

   ANS:
   1.36 A

   DIF: IIIB   OBJ: 19-1.2

7. The amount of charge that moves through the heating coils of a hair dryer in 50.0 s is 37.0 C. What is the current in the hair dryer?

ANS:
0.74 A

DIF: IIIB   OBJ: 19-1.2

8. How much current is in a resistor of 1.8 kΩ if the potential difference across the resistor is 4.0×10^2 V?

ANS:
0.22 A

DIF: IIIA   OBJ: 19-2.1

9. How much current is in a resistor of 280 Ω if there is a potential difference of 120 V across the resistor?

ANS:
0.42 A

DIF: IIIA   OBJ: 19-2.1

10. A resistor has a resistance of 1.8 Ω. How much current is in the resistor if there is a potential difference of 3.0 V across the resistor?

ANS:
1.7 A

DIF: IIIA   OBJ: 19-2.1

11. A resistor has a resistance of 20.0 Ω. How much current is in the resistor if there is a potential difference of 1.5 V across the resistor?

ANS:
0.075 A

DIF: IIIA   OBJ: 19-2.1

12. A potential difference of 12 V is placed across a resistor with 0.25 A of current in it. What is the resistance of the resistor?

ANS:
48 Ω

DIF: IIIA   OBJ: 19-2.1

13. A potential difference of 4.0 V is placed across a resistor with 10.0 A of current in it. What is the resistance of the resistor?

    ANS:
    0.40 Ω

    DIF: IIIA   OBJ: 19-2.1

14. A potential difference of 20.0 V is placed across a resistor with 8.0 A of current in it. What is the resistance of the resistor?

    ANS:
    2.5 Ω

    DIF: IIIA   OBJ: 19-2.1

15. A potential difference of 13 V is placed across a resistor with 1.4 A of current in it. What is the resistance of the resistor?

    ANS:
    9.2 Ω

    DIF: IIIA   OBJ: 19-2.1

16. A 180 Ω resistor has 0.10 A of current in it. What is the potential difference across the resistor?

    ANS:
    18 V

    DIF: IIIA   OBJ: 19-2.1

17. A 4.0 Ω resistor has 0.9 A of current in it. What is the potential difference across the resistor?

    ANS:
    3.6 V

    DIF: IIIA   OBJ: 19-2.1

18. A 13 Ω resistor has 0.05 A of current in it. What is the potential difference across the resistor?

    ANS:
    0.65 V

    DIF: IIIA   OBJ: 19-2.1

19. A 2.0 kΩ resistor has 0.042 A of current in it. What is the potential difference across the resistor?

ANS:
84 V

DIF: IIIA   OBJ: 19-2.1

20. A 5.0 Ω resistor has 0.01 A of current in it. What is the potential difference across the resistor?

ANS:
0.05 V

DIF: IIIA   OBJ: 19-2.1

21. An electric space heater is connected across a 120 V outlet. If the resistance of the heater is 18 Ω, how much power is dissipated in the form of electromagnetic radiation and heat?

ANS:
0.80 kW

DIF: IIIB   OBJ: 19-3.2

22. A toaster is connected across a 120 V outlet. If the resistance of the toaster is 25 Ω, how much power is dissipated in the form of electromagnetic radiation and heat?

ANS:
0.58 kW

DIF: IIIB   OBJ: 19-3.2

23. A lamp is connected across a 120 V outlet. If the resistance of the lamp is 240 Ω, how much power is dissipated in the form of electromagnetic radiation, heat, and light?

ANS:
6.0×10^1 W

DIF: IIIB   OBJ: 19-3.2

24. A hair dryer is connected across a 120 V outlet. If the resistance of the hair dryer is 144 Ω, how much power is dissipated in the form of electromagnetic radiation and heat?

ANS:
1.00×10^2 W

DIF: IIIB   OBJ: 19-3.2

MULTIPLE CHOICE

1. If the potential difference across the bulb in a camping lantern is 9.0 V, what is the potential difference across the battery used to power it?
   a. 1.0 V
   b. 3.0 V
   c. 9.0 V
   d. 18 V

   ANS: C   DIF: I   OBJ: 20-1.3

2. If the potential difference across a pair of batteries used to power a flashlight is 6.0 V, what is the potential difference across the flashlight bulb?
   a. 3.0 V
   b. 6.0 V
   c. 9.0 V
   d. 12 V

   ANS: B   DIF: II   OBJ: 20-1.3

3. If the batteries in a portable CD player provide a terminal voltage of 12 V, what is the potential difference across the entire player?
   a. 3.0 V
   b. 4.0 V
   c. 6.0 V
   d. 12 V

   ANS: D   DIF: II   OBJ: 20-1.3

4. How does the potential difference across the bulb in a flashlight compare with the terminal voltage of the batteries used to power the flashlight?
   a. The potential difference is greater than the terminal voltage.
   b. The potential difference is less than the terminal voltage.
   c. The potential difference is equal to the terminal voltage.
   d. It cannot be determined unless the internal resistance of the batteries is known.

   ANS: C   DIF: II   OBJ: 20-1.3

5. If a 9.0 V battery is connected to a light bulb, what is the potential difference across the bulb?
   a. 3.0 V
   b. 4.5 V
   c. 9.0 V
   d. 18 V

   ANS: C   DIF: I   OBJ: 20-1.3

6. Three resistors with values of 4.0 Ω, 6.0 Ω, and 8.0 Ω, respectively, are connected in series. What is their equivalent resistance?
   a. 18
   b. 8.0
   c. 6.0
   d. 1.8

   ANS: A   DIF: I   OBJ: 20-2.1

7. The following three appliances are connected in series to a 120 V house circuit: a toaster, 1200 W; a coffee pot, 750 W; and a microwave, 6.0×10^2 W. If all were operated at the same time, what total current would they draw?

   a. 3 A   c. 10 A
   b. 5 A   d. 21 A

   ANS: D   DIF: IIIA   OBJ: 20-2.1

8. Three resistors connected in series carry currents labeled I_1, I_2, and I_3, respectively. Which of the following expresses the total current, I_t, in the system made up of the three resistors in series?

   a. $I_t = I_1 + I_2 + I_3$ c. $I_t = I_1 = I_2 = I_3$
   b. $I_t = (1/I_1 + 1/I_2 + 1/I_3)$ d. $I_t = (1/I_1 + 1/I_2 + 1/I_3)^{-1}$

   ANS: C   DIF: II   OBJ: 20-2.1

9. Three resistors connected in series have voltages labeled ΔV_1, ΔV_2, and ΔV_3. Which of the following expresses the total voltage taken over the three resistors together?

   a. $\Delta V_t = \Delta V_1 + \Delta V_2 + \Delta V_3$ c. $\Delta V_t = \Delta V_1 = \Delta V_2 = \Delta V_3$
   b. $\Delta V_t = (1/\Delta V_1 + 1/\Delta V_2 + 1/\Delta V_3)$ d. $\Delta V_t = (1/\Delta V_1 + 1/\Delta V_2 + 1/\Delta V_3)^{-1}$

   ANS: A   DIF: II   OBJ: 20-2.1

10. Three resistors with values of R_1, R_2, and R_3 are connected in series. Which of the following expresses the total resistance, R_t, of the three resistors?

    a. $R_t = R_1 + R_2 + R_3$ c. $R_t = R_1 = R_2 = R_3$
    b. $R_t = (1/R_1 + 1/R_2 + 1/R_3)$ d. $R_t = (1/R_1 + 1/R_2 + 1/R_3)^{-1}$

    ANS: A   DIF: II   OBJ: 20-2.1

11. Three resistors with values of 3.0 Ω, 6.0 Ω, and 12 Ω are connected in series. What is the equivalent resistance of this combination?

    a. 0.58 Ω   c. 7.0 Ω
    b. 1.7 Ω   d. 21 Ω

    ANS: D   DIF: I   OBJ: 20-2.1

12. Three resistors with values of 4.0 Ω, 6.0 Ω, and 10.0 Ω are connected in parallel. What is their equivalent resistance?

    a. 20.0 Ω   c. 6.0 Ω
    b. 7.3 Ω   d. 1.9 Ω

    ANS: D   DIF: IIIA   OBJ: 20-2.2

13. Three resistors connected in parallel carry currents labeled I_1, I_2, and I_3. Which of the following expresses the total current I_t in the combined system?

    a. $I_t = I_1 + I_2 + I_3$ c. $I_t = I_1 = I_2 = I_3$
    b. $I_t = (1/I_1 + 1/I_2 + 1/I_3)$ d. $I_t = (1/I_1 + 1/I_2 + 1/I_3)^{-1}$

    ANS: A   DIF: I   OBJ: 20-2.2

14. Three resistors connected in parallel have voltages labeled ΔV_1, ΔV_2, and ΔV_3. Which of the following expresses the total voltage across the three resistors?
    a. $\Delta V_t = \Delta V_1 + \Delta V_2 + \Delta V_3$
    b. $\Delta V_t = (1/\Delta V_1 + 1/\Delta V_2 + 1/\Delta V_3)$
    c. $\Delta V_t = \Delta V_1 = \Delta V_2 = \Delta V_3$
    d. $\Delta V_t = (1/\Delta V_1 + 1/\Delta V_2 + 1/\Delta V_3)^{-1}$

    ANS: C   DIF: II   OBJ: 20-2.2

15. Three resistors with values of R_1, R_2, and R_3 are connected in parallel. Which of the following expresses the total resistance, R_t, of the three resistors?
    a. $R_t = R_1 + R_2 + R_3$
    b. $R_t = (1/R_1 + 1/R_2 + 1/R_3)$
    c. $R_t = R_1 = R_2 = R_3$
    d. $R_t = (1/R_1 + 1/R_2 + 1/R_3)^{-1}$

    ANS: D   DIF: II   OBJ: 20-2.2

16. Three resistors with values of 3.0 Ω, 6.0 Ω, and 12 Ω are connected in parallel. What is the equivalent resistance of this combination?
    a. 0.26 Ω
    b. 1.7 Ω
    c. 9.0 Ω
    d. 33 Ω

    ANS: B   DIF: IIIA   OBJ: 20-2.2

17. Two resistors with values of 6.0 Ω and 12 Ω are connected in parallel. This combination is connected in series with a 4.0 Ω resistor. What is the overall resistance of this combination?
    a. 0.50 Ω
    b. 2.0 Ω
    c. 8.0 Ω
    d. 22 Ω

    ANS: C   DIF: IIIA   OBJ: 20-3.1

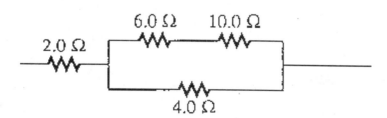

18. What is the equivalent resistance for the resistors in the figure above?
    a. 2.3 Ω
    b. 5.2 Ω
    c. 13 Ω
    d. 22 Ω

    ANS: B   DIF: IIIB   OBJ: 20-3.1

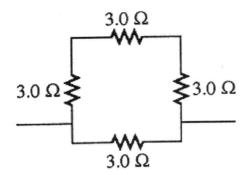

19. What is the equivalent resistance for the resistors in the figure above?
   a.   1.3 Ω
   b.   2.2 Ω
   c.   3.0 Ω
   d.   7.5 Ω

ANS: B   DIF: IIIB   OBJ: 20-3.1

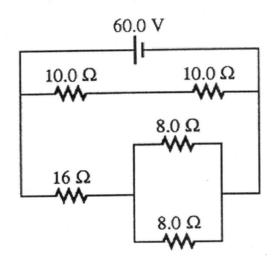

20. What is the equivalent resistance for the resistors in the figure above?
   a.   7.5 Ω
   b. 1.0×10^1 Ω
   c.   16 Ω
   d.   18 Ω

ANS: B   DIF: IIIA   OBJ: 20-3.1

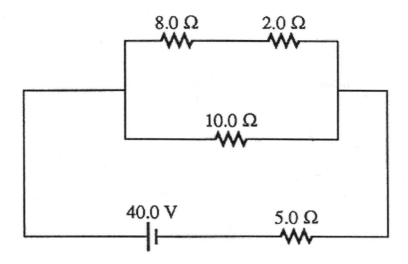

21. What is the equivalent resistance for the resistors in the figure above?
    a. 25 Ω
    b. 1.0×10^1 Ω
    c. 7.5 Ω
    d. 5.0 Ω

    ANS: B   DIF: IIIA   OBJ: 20-3.1

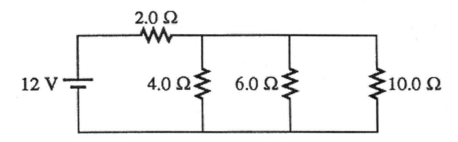

22. Three resistors connected in parallel have individual values of 4.0 Ω, 6.0 Ω, and 10.0 Ω, as shown above. If this combination is connected in series with a 12.0 V battery and a 2.0 Ω resistor, what is the current in the 10.0 Ω resistor?
    a. 0.59 A
    b. 1.0 A
    c. 11A
    d. 16A

    ANS: A   DIF: IIIA   OBJ: 20-3.2

23. Two resistors with values of 6.0 Ω and 12 Ω are connected in parallel. This combination is connected in series with a 2.0 Ω resistor and a 24 V battery. What is the current in the 2.0 Ω resistor?
    a. 2.0 A
    b. 4.0 A
    c. 6.0 A
    d. 12 A

    ANS: A   DIF: IIIA   OBJ: 20-3.2

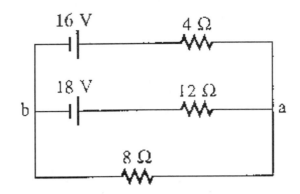

24. What is the current in the 4.0 Ω resistor in the figure above?
    a. 1.0 A   c. 2.5 A
    b. 0.80 A   d. 3.0 A

    ANS: A   DIF: IIIC   OBJ: 20-3.2

25. What is the current in the 8.0 Ω resistor in the figure above?
    a. 1.0 A   c. 1.5 A
    b. 0.50 A   d. 2.0 A

    ANS: C   DIF: IIIC   OBJ: 20-3.2

26. What is the potential difference across point a and point b in the figure above?
    a. 6 V   c. 12 V
    b. 8 V   d. 24 V

    ANS: C   DIF: IIIC   OBJ: 20-3.2

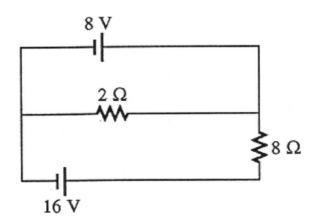

27. What is the current in the 2 Ω resistor in the figure above?
    a. 2 A   c. 4A
    b. 3 A   d. 6A

    ANS: C   DIF: IIIB   OBJ: 20-3.2

28. What is the current in the 8 Ω resistor in the figure above?
    a. 1 A
    b. 2 A
    c. 3 A
    d. 6 A

    ANS: A   DIF: IIIB   OBJ: 20-3.2

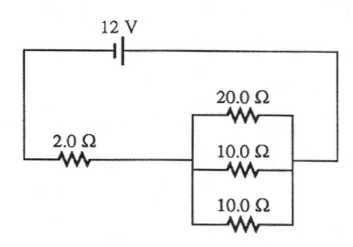

29. How much current is in one of the 10 Ω resistors in the diagram shown above?
    a. 0.8 A
    b. 2 A
    c. 0.6 A
    d. 4 A

    ANS: A   DIF: IIIB   OBJ: 20-3.2

SHORT ANSWER

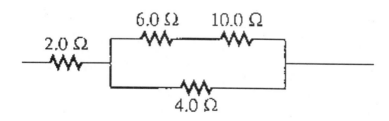

1. Identify the types of elements in the schematic diagram above and the number of each type.

    ANS:
    four resistors

    DIF: I   OBJ: 20-1.1

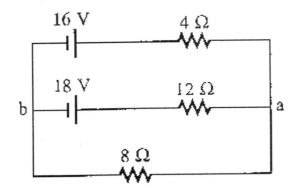

2. Identify the types of elements in the schematic diagram above and the number of each type.

   ANS:
   two batteries, three resistors

   DIF: I   OBJ: 20-1.1

3. Draw a schematic diagram that contains three resistors and one battery.

   ANS:
   okay

   DIF: I   OBJ: 20-1.1

4. Draw a schematic diagram that contains a 1000 V battery, a 3000 Ω resistor, a 0.5 μF capacitor, and an open switch.

   ANS:

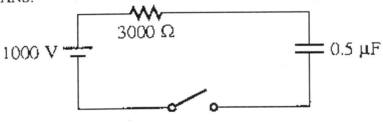

   DIF: I   OBJ: 20-1.1

Holt Physics Assessment Item Listing
214

5. Draw a schematic diagram that contains one battery, two resistors, one capacitor, and one closed switch.

ANS:

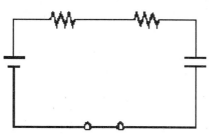

OBJ: 20-1.1

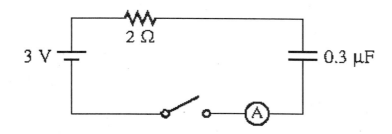

6. In the schematic diagram above, will there be a current?

ANS:
no

DIF: I   OBJ: 20-1.2

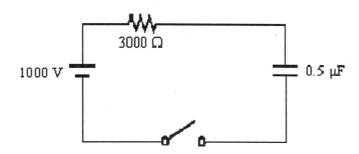

7. In the schematic diagram above, will there be a current?

ANS:
no

DIF: I   OBJ: 20-1.2

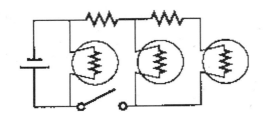

8. Which bulbs will have a current in the schematic diagram above?

ANS:
The first light bulb has a current, but the other two light bulbs do not.

DIF: I   OBJ: 20-1.2

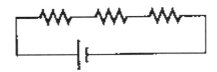

9. In the schematic diagram above, will there be a current?

ANS:
yes

DIF: I   OBJ: 20-1.2

PROBLEM

1. Three resistors with values of 18 Ω, 26 Ω, 9 Ω, respectively, are connected in series. What is their equivalent resistance?

ANS:
53 Ω

DIF: I   OBJ: 20-2.1

2. Three resistors with values of 12 Ω, 38 Ω, 125 Ω, respectively, are connected in series. What is their equivalent resistance?

ANS:
175 Ω

DIF: I   OBJ: 20-2.1

3. Three resistors with values of 28 Ω, 58 Ω, 22 Ω, respectively, are connected in series. What is their equivalent resistance?

ANS:
108 Ω

DIF: I   OBJ: 20-2.1

4. Three resistors with values of 64 Ω, 135 Ω, 92 Ω, respectively, are connected in series. What is their equivalent resistance?

ANS:
291 Ω

DIF: I   OBJ: 20-2.1

5. Three resistors with values of 2 Ω, 9 Ω, 11 Ω, respectively, are connected in series. What is their equivalent resistance?

ANS:
22 Ω

DIF: I   OBJ: 20-2.1

6. Three resistors with values of 16 Ω, 19 Ω, 25 Ω, respectively, are connected in parallel. What is their equivalent resistance?

ANS:
16.4 Ω

DIF: I   OBJ: 20-2.2

7. Three resistors with values of 23 Ω, 81 Ω, 16 Ω, respectively, are connected in parallel. What is their equivalent resistance?

ANS:
8.41 Ω

DIF: I   OBJ: 20-2.2

8. Three resistors with values of 11 Ω, 8 Ω, 2 Ω, respectively, are connected in parallel. What is their equivalent resistance?

ANS:
1.4 Ω

DIF: I   OBJ: 20-2.2

9. Three resistors with values of 9 Ω, 6 Ω, 12 Ω, respectively, are connected in parallel. What is their equivalent resistance?

ANS:
12.8 Ω

DIF: I   OBJ: 20-2.2

10. Three resistors with values of 15 Ω, 41 Ω, 58 Ω, respectively, are connected in parallel. What is their equivalent resistance?

ANS:
9.2 Ω

DIF: I   OBJ: 20-2.2

MULTIPLE CHOICE

1. Where is the magnitude of the magnetic field around a permanent magnet greatest?
   a. close to the poles
   b. far from the poles
   c. The magnitude is equal at all points on the field.
   d. The magnitude depends on the material of the magnet.

   ANS: A

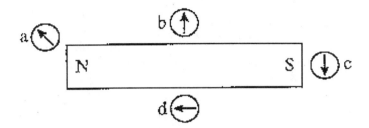

2. Which compass needle orientation in the figure above might correctly describe the magnet's field at that point?
   a. a   c. c
   b. b   d. d

   ANS: A

3. Which describes magnetic declination?
   a. the angle between Earth's magnetic field and Earth's surface
   b. Earth's magnetic field strength at the equator
   c. the tendency for Earth's field to reverse itself
   d. the angle between directions to true north and magnetic

   ANS: D

4. The current in a long wire creates a magnetic field in the region around the wire. How is the strength of that field at a given distance, r, from the wire's center related to r?
   a. the field is directly proportional to r c. the field is directly proportional to r^2
   b. the field is inversely proportional to r d. the field is inversely proportional to r^2

   ANS: B

5. A current in a solenoid coil creates a magnetic field inside the coil. The field strength is directly proportional to the
   a. coil area.   c. coil area and current.
   b. current.   d. length.

   ANS: B

6. A current in a solenoid with *N* turns creates a magnetic field inside the solenoid. The magnetic field strength is directly proportional to the
   a. number of turns only.
   b. strength of the current only.
   c. number of turns in the loop and the strength of the current.
   d. size of the solenoid.

   ANS: C

7. A current in a long, straight wire produces a magnetic field. These magnetic field lines
   a. go out from the wire to infinity.
   b. come in from infinity to the wire.
   c. form circles that pass through the wire.
   d. form circles that go around the wire.

   ANS: D

8. A solenoid is in an upright position on a table. A counterclockwise current of electrons causes the solenoid to have a(n) _____ magnetic pole at its bottom end. If a compass is placed at the top of the solenoid, the north pole of the compass would be
   a. north; attracted
   b. south; attracted
   c. north; repelled
   d. south; repelled

   ANS: A

9. A solenoid is in an upright position on a table. A clockwise current of electrons causes the solenoid to have a(n) _____ magnetic pole at its bottom end. If a compass is placed at the top of the solenoid, the north pole of the compass would be
   a. north; attracted
   b. south; attracted
   c. north; repelled
   d. south; repelled

   ANS: D

10. If a wire is carrying a strong, steady current, the magnetic field is
    a. proportional to the current and inversely proportional to the distance from the wire.
    b. proportional to the current and proportional to the distance from the wire.
    c. inversely proportional to the current and inversely proportional to the distance from the wire.
    d. inversely proportional to the current and proportional to the distance from the wire.

    ANS: A

11. A microscopic magnetic region composed of a group of atoms whose magnetic fields are aligned in a common direction is called a(n) _____. In most materials, when these groups are randomly distributed, the substance will show _____ magnetism.
    a. domain; no
    b. pole; some
    c. cell; unusual
    d. ion; strong

    ANS: A

12. Which orientation characterizes the magnetic domains in a nonmagnetized piece of iron?
    a. parallel to the magnetic axis
    b. antiparallel to the magnetic axis
    c. random
    d. perpendicular to the magnetic axis

    ANS: C

13. In a magnetized substance, the domains
    a. are randomly oriented.
    b. cancel each other.
    c. line up mainly in one direction.
    d. can never be reoriented.

    ANS: C

14. In a permanent magnet,
    a. domain alignment persists after the external magnetic field is removed.
    b. domain alignment becomes random after the external magnetic field is removed.
    c. domains are always randomly oriented.
    d. the magnetic fields of the domains cancel each other.

    ANS: A

15. When an iron rod is inserted into a solenoid coil's center, the magnetic field produced by the current in the loops
    a. causes the iron to return to an unmagnetized state.
    b. forces the domain in the iron out of alignment.
    c. causes random orientation of the domains in the iron.
    d. causes alignment of the domains in the iron.

    ANS: D

16. In soft magnetic materials such as iron, what happens when an external magnetic field is removed?
    a. The domain alignment persists.
    b. The orientation of domains fluctuates.
    c. The material becomes a hard magnetic material.
    d. The orientation of domains changes, and the material returns to an unmagnetized state.

    ANS: D

17. An electron that moves with a speed of 3.0×10^4 m/s perpendicular to a uniform magnetic field of 0.40 T experiences a force of what magnitude? ($e = 1.60 \times 10^{-19}$ C)
    a. 4.8×10^{-14} N
    b. 1.9×10^{15} N
    c. 2.2×10^{24} N
    d. 0

    ANS: B

18. An electron moves across Earth's equator at a speed of 2.5×10^6 m/s and in a direction 35° north of east. At this point, Earth's magnetic field has a direction due north, is parallel to the surface, and has a value of 0.10×10^{-4} T. What is the magnitude of the force acting on the electron due to its interaction with Earth's magnetic field? ($e = 1.60 \times 10^{-19}$ C)
    a. 5.1×10^{-18} N
    b. 4.0×10^{-18} N
    c. 3.3×10^{-18} N
    d. 2.3×10^{-18} N

    ANS: C

19. An electron moves north at a velocity of 4.5×10^4 m/s and has a force of 7.2×10^{-18} N exerted on it. If the magnetic field points upward, what is the magnitude of the magnetic field?
    a. 2.0 mT   c. 3.6 mT
    b. 1.0 mT   d. 4.8 mT

    ANS: B

20. The direction of the force on a current-carrying wire in an external magnetic field is
    a. perpendicular to the current only.
    b. perpendicular to the magnetic field only.
    c. perpendicular to the current and to the magnetic field.
    d. parallel to the current and to the magnetic field.

    ANS: C

21. If a proton is released at the equator and falls toward Earth under the influence of gravity, the magnetic force on the proton will be toward the
    a. north.   c. east.
    b. south.   d. west.

    ANS: C

22. What is the path of an electron moving perpendicular to a uniform magnetic field?
    a. a straight line   c. an ellipse
    b. a circle   d. a parabola

    ANS: B

23. What is the path of an electron moving parallel to a uniform magnetic field?
    a. straight line   c. ellipse
    b. circle   d. parabola

    ANS: A

24. A 2.0 m wire segment carrying a current of 0.60 A oriented parallel to a uniform magnetic field of 0.50 T experiences a force of what magnitude?
    a. 0.60 N   c. 0.15 N
    b. 0.30 N   d. 0.0 N

    ANS: D

25. A stationary positive charge, Q, is located in a magnetic field, B, which is directed toward the right, as shown in the figure above. The direction of the magnetic force on Q is
    a. toward the right.   c. down.
    b. up.   d. There is no magnetic force.

    ANS: D

26. A current-carrying wire 0.50 m long is positioned perpendicular to a uniform magnetic field. If the current is 10.0 A and there is a resultant force of 3.0 N on the wire due to the interaction of the current and field, what is the magnetic field strength?
   a. 0.60 T
   b. 15 T
   c. 1.8×10^3 T
   d. 6.7×10^3 T

   ANS: A

27. Consider two long, straight, parallel wires, each carrying a current I. If the currents move in opposite directions,
   a. the two wires will attract each other.
   b. the two wires will repel each other.
   c. the two wires will exert a torque on each other.
   d. neither wire will exert a force on the other.

   ANS: B

28. Consider two long, straight, parallel wires, each carrying a current I. If the currents move in the same direction,
   a. the two wires will attract each other.
   b. the two wires will repel each other.
   c. the two wires will exert a torque on each other.
   d. neither wire will exert a force on the other.

   ANS: A

29. A current-carrying conductor in and perpendicular to a magnetic field experiences a force that is
   a. perpendicular to the current.
   b. parallel to the current.
   c. inversely proportional to the potential difference.
   d. inversely proportional to the velocity

   ANS: A

SHORT ANSWER

1. A bar magnet is suspended and allowed to rotate freely. If the magnetic field of Earth is considered to be equivalent to that of a large bar magnet, which pole of the suspended magnet would point toward the magnetic north pole of Earth?

   ANS:
   the south (S) pole   DIF II

2. If the head of an iron nail touches a magnet, the nail will become a magnet by induction. If the nail touches the north pole of the magnet, what kind of pole is at the point of the nail? Explain.

   ANS:
   The end of the magnetized nail touching the north pole of the magnet will be a south pole by induction. Otherwise, it would be repelled by the magnet. The tip of the nail that points away from the magnet must have the opposite polarity and thus will be a north pole.

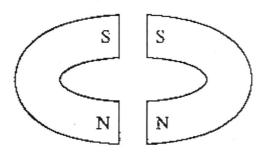

3. Will the magnets in the figure above attract or repel each other?

ANS:
repel

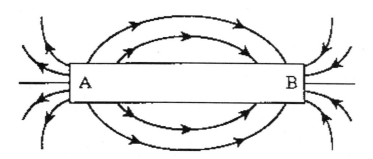

4. Will the magnets in the figure above attract or repel each other?

ANS:
repel

5. How can the magnetic field around a permanent magnet be determined?

ANS:
It can be determined with a compass. The direction of the magnetic field is defined as the direction in which the north pole of a compass needle points at that location.

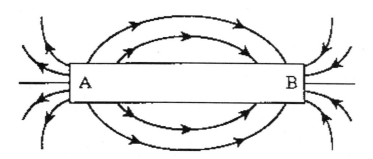

6. The magnetic field of a bar magnet is shown in the figure above. Is the magnet's north pole at A or B?

ANS:
A

7. Which magnetic pole is at the geographic South Pole of Earth?

ANS:
The magnetic north pole is located at the geographic South Pole of Earth.

8. Which magnetic pole is at the geographic North Pole of Earth?

ANS:
The magnetic south pole is located at the geographic North Pole of Earth.

9. At points near Earth's equator, one would expect the magnetic inclination angle to be approximately _____.

ANS:
$0°$

10. Describe how the right-hand rule applies to the magnetic field produced by the current in a straight conductor.

ANS:
If a wire is grasped in the right hand with the thumb pointing in the direction of the current, the fingers will curl in the direction of the magnetic field.

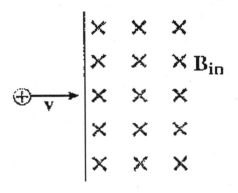

11. Find the direction of the force on an electron moving through the magnetic field shown above.

ANS:
down, toward the bottom of the page

12. Find the direction of the force on an electron moving through the magnetic field shown above.

ANS:
up, toward the top of the page

13. Find the direction of the force on an electron moving through the magnetic field shown above.

ANS:
into the page.

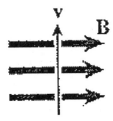

14. Find the direction of the force on an electron moving through the magnetic field shown above.

    ANS:
    out of the page

15. Find the direction of the force on an electron moving through the magnetic field shown above.

    ANS:
    to the right

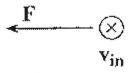

16. A negative charge is moving through a magnetic field. The direction of motion and the direction of the force acting on it at one moment are shown in the figure above. Find the direction of the magnetic field.

    ANS:
    up, toward the top of the page

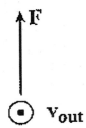

F

(•) **v**_{out}

17. A negative charge is moving through a magnetic field. The direction of motion and the direction of the force acting on it at one moment are shown in the figure. Find the direction of the magnetic field.

    ANS:
    to the left

18. A negative charge is moving through a magnetic field. The direction of motion and the direction of the force acting on it at one moment are shown in the figure. Find the direction of the magnetic field.

    ANS:
    out of the page

19. Electrons move from the south to the north in a wire. What is the direction of the magnetic field at a point directly above the wire?

    ANS:
    The magnetic field points to the west.

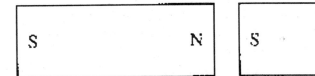

20. Will the magnets in the figure above attract or repel each other?

    ANS:
    attract

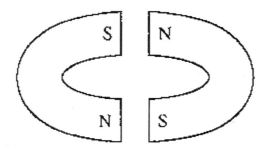

21. Will the magnets in the figure above attract or repel each other?

ANS:
attract

PROBLEM

1. An electron moves north at a velocity of 9.8×10^4 m/s and has a magnetic force of 5.6×10^{-18} N exerted on it. If the magnetic field points upward, what is the magnitude of the magnetic field?

ANS:
3.6×10^{-4} T

2. An electron moves north at a velocity of 8.4×10^4 m/s and has a magnetic force of 3.0×10^{-18} N exerted on it. If the magnetic field points upward, what is the magnitude of the magnetic field?

ANS:
2.2×10^{-4} T

3. An electron moves north at a velocity of 7.3×10^4 m/s and has a magnetic force of 1.8×10^{-18} N exerted on it. If the magnetic field points upward, what is the magnitude of the magnetic field?

ANS:
1.54×10^{-4} T

4. An electron moves north at a velocity of 6.2×10^6 m/s and has a magnetic force of 2.1×10^{-18} N exerted on it. If the magnetic field points upward, what is the magnitude of the magnetic field?

ANS:
2.1×10^{-6} T

Holt Physics Assessment Item Listing
228

5. An electron moves north at a velocity of 5.8×10^5 m/s and has a magnetic force of 3.7×10^{-18} N exerted on it. If the magnetic field points upward, what is the magnitude of the magnetic field?

ANS:
4.0×10^{-5} T

6. An electron moves north at a velocity of 2.7×10^4 m/s and has a magnetic force of 9.5×10^{-18} N exerted on it. If the magnetic field points upward, what is the magnitude of the magnetic field?

ANS:
2.2×10^{-3} T

7. An electron moves north at a velocity of 1.6×10^4 m/s and has a magnetic force of 1.6×10^{-18} N exerted on it. If the magnetic field points upward, what is the magnitude of the magnetic field?

ANS:
6.2×10^{-4} T

8. An electron moves north at a velocity of 9.1×10^5 m/s and has a magnetic force of 3.2×10^{-18} N west exerted on it. If the magnetic field points upward, what is the magnitude of the magnetic field?

ANS:
2.2×10^{-5} T

9. An electron moves north at a velocity of 4.9×10^6 m/s and has a magnetic force of 7.5×10^{-18} N west exerted on it. If the magnetic field points upward, what is the magnitude of the magnetic field?

ANS:
9.6×10^{-6} T

10. An electron moves north at a velocity of 3.8×10^5 m/s and has a magnetic force of 8.9×10^{-18} N west exerted on it. If the magnetic field points upward, what is the magnitude of the magnetic field?

ANS:
1.4×10^{-4} T

MULTIPLE CHOICE

1. A loop of wire is held in a vertical position at the equator with the face of the loop facing in the east-west direction. What change will induce the greatest current in the loop?
    a. raising the loop to a higher elevation
    c. rotating the loop so its face is vertical
    b. moving the loop north
    d. rotating the loop so its face is north-south

    ANS: D

2. A bar magnet falls through a loop of wire with constant velocity, and the north pole enters the loop first. The induced current will be greatest when the magnet is located so that the loop is
    a. near either the north or the south pole.
    b. near the north pole only.
    c. near the middle of the magnet.
    d. With no acceleration, the induced current is zero.

    ANS: A

3. Electricity may be generated by rotating a loop of wire between the poles of a magnet. The induced current is greatest when
    a. the plane of the loop is parallel to the magnetic field.
    b. the plane of the loop is perpendicular to the magnetic field.
    c. the magnetic flux through the loop is a maximum.
    d. the plane of the loop makes an angle of 45° with the magnetic field.

    ANS: B

4. A loop of wire is rotated 360° across an external magnetic field. During one period of revolution, the induced current changes in
    a. magnitude only.
    c. both magnitude and direction.
    b. direction only.
    d. amplitude

    ANS: C

5. According to Lenz's law, the magnetic field of an induced current in a conductor will
    a. enhance the applied field.
    b. heat the conductor.
    c. increase the potential difference.
    d. oppose a change in the applied magnetic field.

    ANS: D

6. Which statement is correct?
    a. The magnet field of an induced current opposes the applied magnetic field.
    b. According to the principle of energy conservation, an induced field attempts to keep the total field strength constant.
    c. An induced electric field opposes an applied magnetic field.
    d. Lenz's law is used to find the average induced emf.

    ANS: B

Holt Physics Assessment Item Listing
230

7. According to Lenz's law, if the applied magnetic field changes,
   a. the induced field attempts to keep the total field strength constant.
   b. the induced field attempts to increase the total field strength.
   c. the induced field attempts to decrease the total field strength.
   d. the induced field attempts to oscillate about an equilibrium value.

   ANS: A

8. A circular coil with an area of 5.0×10^2 m^2 and with 500 turns of wire is placed in a uniform magnetic field perpendicular to the plane of the coil. If the field changes in value from -0.100 T to $+0.150$ T in an interval of 0.500 s, what is the induced emf in the coil?
   a. -12.5 V
   b. 125 V
   c. 188 V
   d. 252 V

   ANS: A

9. A coil with 25 turns of wire moves in a uniform magnetic field of 1.5 T. The plane of the magnetic field is perpendicular to the plane of the coil. The coil has a cross-sectional area of 0.80 m^2. The coil exits the field in 1.0 s. If the coil's resistance is 2.0 Ω, what is the induced current?
   a. 5.0 A
   b. 15 A
   c. 2.0×10^1 A
   d. 25 A

   ANS: A

10. A coil with a wire that is wound around a 2.0 m^2 hollow tube 35 times. A uniform magnetic field is applied perpendicular to the plane of the coil. If the field changes uniformly from 0.00 T to 0.55 T in 0.85 s, what is the induced emf in the coil?
    a. -33 V
    b. 33 V
    c. -45 V
    d. 45 V

    ANS: C

11. A coil with a wire that is wrapped 40.0 turns around a 0.5 m^2 hollow tube. A uniform magnetic field is applied perpendicular to the plane of the coil. The field changes uniformly from 0.00 T to 0.95 T in 2.0 s. If the resistance in the coil is 3.0 Ω, what is the magnitude of the induced current in the coil?
    a. 11 A
    b. -11 A
    c. 9.5 A
    d. -9.5 A

    ANS: D

12. A 500-turn circular coil with an area of 0.055 m^2 is mounted on a rotating frame that turns at a constant frequency of 60.0 Hz in a magnetic field of 0.15 T. What is the maximum induced emf in the coil?
    a. 1.6×10^3 V
    b. 2.2×10^2 V
    c. 7.9×10^2 V
    d. 2.5×10^3 V

    ANS: A

13. A generator consists of eight turns of wire, each turn has an area of 0.15 m². The loop rotates in a magnetic field of 0.55 T at a constant frequency of 9.0 Hz. What is the maximum induced emf in the loop?
   a. 19 V   c. 5.9 V
   b. 37 V   d. 4.1 V

   ANS: B

14. A maximum emf of 1.6×10^2 V is induced in a generator coil rotating with a frequency of 60.0 Hz. If the coil has an area of 0.095 m² and rotates in a magnetic field of 0.55 T, how many turns are in the coil?
   a. 4   c. 8
   b. 6   d. 16

   ANS: C

15. A circular coil with a radius of 0.33 m and 15 turns rotates at a constant frequency of 3.0 Hz in a uniform magnetic field of 1.5 T. What is the maximum emf induced in the coil?
   a. 8.8×10^2 V   c. 2.9×10^2 V
   b. 3.4×10^2 V   d. 1.5×10^2 V

   ANS: D

16. A generator consists of 10.0 turns of wire, each turn has an area of 0.095 m². If a maximum emf of 1.2×10^2 V is induced when the coil rotates with a frequency of 60.0 Hz, what is the strength of the magnetic field?
   a. 0.33 T   c. 1.50 T
   b. 0.15 T   d. 0.85 T

   ANS: A

17. A generator supplies an rms current of 1.66 A. If the resistance of the circuit is 66.0 Ω, what is the maximum emf?
   a. 77.5 V   c. 125 V
   b. 155 V   d. 38.7 V

   ANS: B

18. The maximum current supplied by a generator to a 25 Ω circuit is 6.2 A. What is the rms potential difference?
   a. 150 V   c. 110 V
   b. 120 V   d. 62 V

   ANS: C

19. An ac generator has a maximum output emf of 215 V. What is the rms potential difference?
   a. 145 V   c. 216 V
   b. 152 V   d. 304 V

   ANS: B

20. An ac generator has a maximum output emf of 120 V. The generator is connected to a 125 Ω resistor. What is the rms current through the resistor?

    a.   0.68 A   c.   1.73 A

    b.   0.67 A   d.   2.43 A

ANS: A

21. An ac generator has a maximum emf output of 150 V. What is the rms current in the circuit when the generator is connected to a 35 Ω resistor?

    a.   3.0 A   c.   2.6 A

    b.   1.5 A   d.   1.2 A

ANS: A

22. The operation of an electric motor depends on

    a.   the Doppler effect.

    b.   the photoelectric effect.

    c.   the force acting on a current-carrying wire in a magnetic field.

    d.   induced commutators from the rotation of a wire in a magnetic field.

ANS: C

23. Which conversion process is the basic function of the electric generator?

    a.   mechanical energy to electrical energy   c.   low voltage to high voltage, or vice versa

    b.   electrical energy to mechanical energy   d.   alternating current to direct current

ANS: A

24. Which conversion process is the basic function of the electric motor?

    a.   mechanical energy to electrical energy   c.   low voltage to high voltage, or vice versa

    b.   electrical energy to mechanical energy   d.   alternating current to direct current

ANS: B

25. Under which condition is the back emf in an electric motor at its maximum value?

    a.   motor speed is zero   c.   voltage is at maximum

    b.   current is at maximum   d.   motor speed is at maximum

ANS: D

26. In most electric generators, either the armature or the magnetic field is _____, generating a(n) _____.

    a.   rotated; induced current   c.   interrupted; impulse change

    b.   turned off; temporary dipole   d.   nonconducting; flux line

ANS: A

27. In a primary-secondary coil combination, the current in the secondary is maximum at the moment that which condition is met in the primary?
    a. the current is maximum in positive direction
    b. the current is maximum in negative direction
    c. the rate of current change is maximum
    d. the voltage is maximum in a positive direction

    ANS: C

28. Two loops of wire are arranged so that a changing current in the primary will induce a current in the secondary. The secondary loop has twice as many turns as the primary loop. As long as the current in the primary is steady at 3.0 A, the current in the secondary will be
    a. 3.0 A.   c. 1.5 A.
    b. 6.0 A.   d. zero.

    ANS: D

29. In a two-coil system, the mutual inductance depends on
    a. only the geometrical properties of the coils.
    b. only the orientation of the coils to each other.
    c. both the geometrical properties of the coils and their orientation to each other.
    d. neither the geometrical properties of the coils nor their orientation to each other.

    ANS: C

30. In a two-coil system, the induced potential difference in the secondary coil depends on
    a. the orientation of the coils being kept constant.
    b. the number of turns of wire.
    c. the switch being kept open.
    d. the iron ring around which the coils are wrapped.

    ANS: B

31. A step-up transformer used on a 120 V line has 95 turns on the primary and 2850 turns on the secondary. What is the potential difference across the secondary?
    a. 30 V   c. 2400 V
    b. 1800 V   d. 3600 V

    ANS: D

32. A step-down transformer has 2500 turns on its primary and 5.0×10^1 turns on its secondary. If the potential difference across the primary is 4850 V, what is the potential difference across the secondary?
    a. 1.0 V   c. 97 V
    b. 110 V   d. 240 V

    ANS: C

33. A potential difference of 115 V across the primary of a step-down transformer provides a potential difference of 2.3 V across the secondary. What is the ratio of the number of turns of wire on the primary to the number of turns on the secondary?
    a. 1:50
    a. A   c.   25:1
    b. 50:1   d.   1:25

    ANS: B

34. A transformer has 15 turns in its primary and 6750 turns in its secondary. If the potential difference across the primary is 1.2 V, what is the potential difference across the secondary?
    a. 540 V   c.   270 V
    b. 380 V   d.   750 V

    ANS: A

35. The self-inductance of a solenoid increases under which condition?
    a. solenoid length is increased
    b. cross-sectional area is decreased
    c. number of coils per unit length is decreased
    d. number of coils is increased

    ANS: D

36. How is the emf in a current-carrying inductor related to its self-inductance, L?
    a. It is directly proportional to L^2.   c.   It is directly proportional to L.
    b. It is directly proportional to $L^{1/2}$. d.   It is inversely proportional to L.

    ANS: C

37. How is the energy stored in a current-carrying inductor related to the current value, I?
    a. It is directly proportional to I^2.   c.   It is directly proportional to $I^{1/2}$.
    b. It is directly proportional to ΔI. d.   It is inversely proportional to ΔI.

    ANS: B

38. Self-induction occurs when the changing current in a circuit
    a. induces a current in a nearby circuit.
    b. induces a voltage opposing the current change.
    c. changes the potential difference of a direct current.
    d. decreases the magnetic field of the original current.

    ANS: B

39. By what factor is the self-induction of a coil changed if its number of coil turns is tripled?
    a. $\frac{1}{3}$ c.   6
    b. 3   d.   9

    ANS: B

Holt Physics Assessment Item Listing
235

SHORT ANSWER

1. List three ways to induce a current in a circuit loop using only a magnet.
    1.
    2.
    3.

    ANS:
    1. Move the circuit loop into or out of the magnetic field. 2. Rotate the circuit loop in the magnetic field so that the angle between the plane of the circuit loop and magnetic field changes. 3. Vary the intensity of the magnetic field.

2. List three essential components of a generator.

    ANS:
    field magnet, slip rings, and brushes

PROBLEM

1. An ac generator has a maximum output emf of 4.2×10^2 V. What is the rms potential difference?

    ANS:
    297 V

2. An AC generator has a maximum output emf of 6.9×10^2 V. What is the rms potential difference?

    ANS:
    488 V

3. An ac generator has a maximum output emf of 1.2×10^3 V. What is the rms potential

    ANS:
    848 V

4. An ac generator has a maximum output emf of 567 V. What is the rms potential difference?

    ANS:
    401 V

5. An ac generator has a maximum output emf of 72.0 V. What is the rms potential difference?

    ANS:
    51 V

6. A step-up transformer used on a 120 V line has 38 turns on the primary and 5163 turns on the secondary. What is the potential difference across the secondary?

   ANS:
   1.6×10^4 V

7. A step-up transformer used on a 120 V line has 49 turns on the primary and 4102 turns on the secondary. What is the potential difference across the secondary?

   ANS:
   1.0×10^4 V

8. A step-up transformer used on a 120 V line has 56 turns on the primary and 1027 turns on the secondary. What is the potential difference across the secondary

   ANS:
   2.2×10^3 V

9. A step-up transformer used on a 120 V line has 71 turns on the primary and 3684 turns on the secondary. What is the potential difference across the secondary?

   ANS:
   6.2×10^3 V

10. A step-up transformer used on a 120 V line has 19 turns on the primary and 9691 turns on the secondary. What is the potential difference across the secondary?

    ANS:
    6.1×10^4 V

MULTIPLE CHOICE

1. What is the energy associated with a light quantum that has a wavelength of 5.0×10^{-7} m? ($h = 6.63 \times 10^{-34}$ J•s)
    a. 4.0×10^{-19} J
    b. 3.3×10^{-19} J
    c. 1.5×10^{-19} J
    d. 1.7×10^{-19} J

    ANS: A

2. What is the frequency of a photon with an energy of 1.7×10^{-19} J? ($h = 6.63 \times 10^{-34}$ J•s)
    a. 3.2×10^{14} Hz
    b. 4.4×10^{14} Hz
    c. 2.6×10^{14} Hz
    d. 6.8×10^{14} Hz

    ANS: C

3. What is the energy of a photon whose frequency is 6.0×10^{20} Hz? ($h = 6.63 \times 10^{-34}$ J•s; 1 eV $= 1.60 \times 10^{19}$ J)
    a. 1.6 MeV
    b. 2.5 MeV
    c. 3.3 MeV
    d. 4.8 MeV

    ANS: B

4. A quantum of radiation has an energy of 2 keV. What is its frequency? ($h = 6.63 \times 10^{-34}$ J•s; 1 eV $= 1.60 \times 10^{-19}$ J)
    a. 3.2×10^{17} Hz
    b. 4.8×10^{17} Hz
    c. 6.3×10^{17} Hz
    d. 7.3×10^{17} Hz

    ANS: B

5. How much energy does a photon of red light that has a wavelength of 640 nm contain? ($h = 6.63 \times 10^{-34}$ J•s; 1 eV $= 1.6 \times 10^{-19}$ J)
    a. 3.2 eV
    b. 2.5 eV
    c. 1.9 eV
    d. 1.3 eV

    ANS: C

6. A monochromatic light beam with a quantum energy value of 3.0 eV is incident upon a photocell. The work function of the target metal is 1.60 eV. What is the maximum kinetic energy of ejected electrons?
    a. 4.6 eV
    b. 4.8 eV
    c. 1.4 eV
    d. 2.4 eV

    ANS: C

7. If barium has a work function of 2.60 eV, what would the wavelength be when $KE_{max} = 0$?
    ($h = 6.63 \times 10^{-34}$ J•s; $c = 3.00 \times 10^8$ m/s; 1 eV $= 1.60 \times 10^{-19}$ J)
    a.   398 nm   c.   497 nm
    b.   478 nm   d.   596 nm

    ANS: B

8. Blue light with a wavelength of 460 nm is incident on a piece of potassium. The work function of
    potassium is 2.2 eV. What is the maximum kinetic energy of the ejected photoelectrons?
    ($h = 6.63 \times 10^{-34}$ J•s; $c = 3.00 \times 10^8$ m/s; 1 eV $= 1.60 \times 10^{-19}$ J)
    a.   1.0 eV   c.   0.25 eV
    b.   0.50 eV   d.   4.9 eV

    ANS: B

9. Light with a wavelength of 350 nm falls on a potassium surface. The work function of potassium
    is 2.24 eV. What is the maximum kinetic energy of the photoelectrons?
    a.   1.2 eV   c.   2.6 eV
    b.   1.3 eV   d.   3.5 eV

    ANS: B

10. A sodium surface is illuminated with light that has a frequency of 1.00×10^{15} Hz. The threshold
    frequency of sodium is 5.51×10^{14} Hz. The maximum kinetic energy of the photoelectrons is
    1.86 eV. What is the work function of sodium?
    a.   1.90 eV   c.   2.28 eV
    b.   2.08 eV   d.   3.23 eV

    ANS: C

11. According to the Rutherford model of the atom, most of the volume of an atom
    a.   is empty space.   c.   contains positive charges
    b.   is occupied by the nucleus.   d.   excludes electrons.

    ANS: A

12. According to the Rutherford model of the atom, what is the concentration of positive charge and
    mass?
    a.   alpha particle   c.   positron
    b.   radius   d.   nucleus

    ANS: D

13. Which statement about Rutherford's model of the atom is NOT correct?
    a.   The model states that positive charge is unevenly distributed.
    b.   The model shows that electrons orbit the nucleus as planets orbit the sun.
    c.   The model explains spectral lines.
    d.   The model states that atoms are unstable.

    ANS: C

14. When a high voltage is applied to a low-pressure gas, causing it to glow, it will emit what type of spectrum?
    a. line emission
    b. line absorption
    c. continuous
    d. monochromatic

    ANS: A

15. Which statement about line-emission spectra is correct?
    a. All the lines are evenly spaced.
    b. All noble gases have the same spectra.
    c. Only gases emit emission lines.
    d. All lines result from discrete energy differences.

    ANS: D

16. When light from argon gas is passed through a prism,
    a. a series of discrete bright lines is observed.
    b. a continuous spectrum is observed.
    c. a series of dark lines imposed on a continuous spectrum is observed.
    d. an absorption spectrum is produced.

    ANS: A

17. The wavelengths contained in an emission spectrum
    a. are the same for all elements.
    b. are characteristic of the element emitting the light.
    c. are the same for all atomic gases.
    d. form dark lines on a continuous spectrum.

    ANS: B

18. The bright lines in the absorption spectrum of an element can be accounted for by the
    a. absorption of photons that occurs when electrons jump from a higher-energy state to a lower-energy state.
    b. emission of photons that occurs when electrons jump from a higher-energy state to a lower-energy state.
    c. absorption of photons that occurs when electrons jump from a lower-energy state to a higher-energy state.
    d. emission of photons that occurs when electrons jump from a lower-energy state to a higher-energy state.

    ANS: B

19. The dark lines in the absorption spectrum of an element can be accounted for by the
   a. absorption of photons that occurs when electrons jump from a higher-energy state to a lower-energy state.
   b. emission of photons that occurs when electrons jump from a higher-energy state to a lower-energy state.
   c. absorption of photons that occurs when electrons jump from a lower-energy state to a higher-energy state.
   d. emission of photons that occurs when electrons jump from a lower-energy state to a higher-energy state.

   ANS: C

20. The probability that an electron will return to a lower energy level by emitting a photon is called
   a. spontaneous emission.
   b. line emission.
   c. the Bohr radius.
   d. the ground state.

   ANS: C

21. The wave-particle duality of light means that in a single experiment
   a. light will act both like a wave and like a particle.
   b. light will act either like a wave or like a particle.
   c. light will not act like a wave nor like a particle.
   d. light always exists as two waves or as two particles.

   ANS: B

22. What is the de Broglie wavelength for a proton that has a mass of 1.67×10^{-27} kg and is moving at a speed of 5.0×10^{-5} m/s? ($h = 6.63 \times 10^{-34}$ J•s)
   a. 1.1×10^{12} m
   b. 4.2×10^{-13} m
   c. 1.8×10^{12} m
   d. 7.9×10^{-13} m

   ANS: D

23. What is the de Broglie wavelength for a proton that has a mass of 1.67×10^{-27} kg and is moving at a speed of 6.0×10^{7} m/s? ($h = 6.63 \times 10^{-34}$ J•s)
   a. 1.5×10^{14} m
   b. 4.8×10^{-11} m
   c. 3.0×10^{8} m
   d. 6.6×10^{-15} m

   ANS: D

24. What speed does a 0.06 kg golf ball have if its de Broglie wavelength is 4.25×10^{-34} m?
   a. 15 m/s
   b. 26 m/s
   c. 31 m/s
   d. 48 m/s

   ANS: B

25. What is the speed of a 50 g rock if its de Broglie wavelength is 3.32×10^{-34} m? ($h = 6.63 \times 10^{-34}$ J•s)
   a. 40 m/s
   b. 30 m/s
   c. 2O m/s
   d. 6O m/s

   ANS: A

26. According to de Broglie, as the momentum of a moving particle is tripled, the corresponding wavelength changes by what factor?

    a. $\frac{1}{9}$ c. 3

    b. $\frac{1}{3}$ d. 9

    ANS: B

27. According to Heisenberg, as soon as the momentum of an electron is known, its
    a. position is known with exact precision.   c. position becomes more certain.
    b. energy becomes stabilized.   d. wavelength becomes more uncertain.

    ANS: C

28. In order to find the momentum of an electron more precisely, you need to
    a. use a lower-energy photon.   c. use a high-frequency photon.
    b. use a higher-energy photon.   d. use a short-wavelength photon.

    ANS: A

29. According to Heisenberg,
    a. measurements in experiments are completely accurate.
    b. it is easy to simultaneously measure the position and momentum of an object with complete certainty.
    c. the measurement procedure limits the accuracy with which the position and momentum of an object can be simultaneously determined.
    d. experiments can be designed to simultaneously reveal the particle and wave properties of light.

    ANS: C

30. One result of the dual nature of light is that
    a. there is no limitation to the accuracy of measurements in experiments.
    b. there are limitations to the accuracy of measurements in experiments.
    c. the principles of classical mechanics hold true.
    d. the uncertainty principle is invalid.

    ANS: B

31. According to Heisenberg, as soon as the exact location of an electron is known, its
    a. exact momentum is known.   c. momentum becomes uncertain.
    b. energy becomes stabilized.   d. direction becomes stabilized.

    ANS: C

32. The quantum-mechanical model of the hydrogen atom suggests a picture of the electron as
    a. a raisin in pudding.   c. a planetary orbiting body.
    b. a probability cloud.   d. a light quantum.

    ANS: B

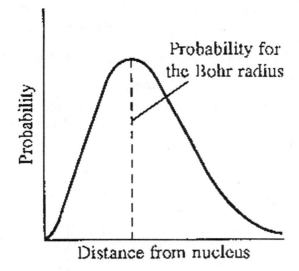

Probability for the Bohr radius

Probability

Distance from nucleus

33. The peak of the curve in the figure above represents the distance from the nucleus of a hydrogen atom at which
   a. there is zero probability of finding the electron.
   b. the electron's location can be precisely determined.
   c. Heisenberg's uncertainty principle is violated.
   d. the electron is most likely to be found in the ground state.

   ANS: D

34. An electron's location can be described by a probability wave because
   a. the electron's location can be precisely determined.
   b. electrons violate Heisenberg's uncertainty principle.
   c. the electron may be found at various distances from the nucleus.
   d. the probability of finding the electron at a distance corresponding to the first Bohr orbit is less than that of finding it at any other distance.

   ANS: C

35. The density of an electron cloud
   a. is proportional to the probability of finding the electron at various locations.
   b. is consistent throughout because the electron could be located at any point within the cloud.
   c. shows that the position of the electron is confined to points in a plane.
   d. accounts for all aspects of Bohr's model of the atom.

   ANS: A

SHORT ANSWER

1. What was Planck's radical assumption about resonators in relation to blackbody radiation?

    ANS:
    Planck proposed that billions of resonators, which are submicroscopic electric oscillators, lined the walls of a glowing cavity. He believed these resonators could only absorb and then reemit discrete amounts of light energy called quanta.

2. How did Planck resolve the ultraviolet catastrophe in blackbody radiation?

    ANS:
    Planck used a mathematical technique instead of classical physics to make calculations. He assumed that discrete amounts of energy were infinitesimally small. Eventually, he derived a mathematical equation reflecting for the quantization of energy.

3. What does the equation $E = hf$ represent?

    ANS:
    The energy of a light quantum is equivalent to the difference in energy between two adjacent energy levels.

4. According to quantum mechanics, when will a resonator radiate or absorb energy? What happens to a resonator that remains in one quantum state?

    ANS:
    When a resonator changes quantum states, it will radiate or absorb energy. While it remains in one quantum state, no energy is absorbed or emitted

5. How did Rutherford explain why electrons in the outer region of the atom are not pulled into the nucleus?

    ANS:
    Rutherford proposed that electrons move in orbits around the nucleus like the planets orbit the sun.

6. .   What are some weaknesses of Rutherford's theory?

    ANS:
    It predicts that atoms are unstable, and it does not explain spectral lines.

7. What is an emission spectrum?

    ANS:
    An emission spectrum is a unique series of spectral lines emitted by anatomic gas when a potential difference is applied across the gas.

Holt Physics Assessment Item Listing
244

8. What is an absorption spectrum?

ANS:
An absorption spectrum is a continuous spectrum interrupted by dark lines that are characteristic of the medium through which the radiation has passed.

9. How can an absorption spectrum be obtained?

ANS:
An absorption spectrum can be obtained by passing light that contains all wavelengths through a vapor of the element being analyzed.

10. What is unique about Bohr's model of the atom?

ANS:
According to Bohr's model, only certain orbits are stable, and the electron is never found between these orbits

11. Is light a particle or a wave? Explain.

ANS:
Light is considered to be both a particle and a wave. Some experiments support the particle theory, and others support the wave theory.

12. Which model of light best explains the interference phenomena?

ANS:
the wave model

13. Which model of light best explains photoelectrons?

ANS:
the particle model

14. In quantum mechanics, an electron's location is described by a(n)

ANS:
probability wave.

PROBLEM

1. What is the energy of a photon whose frequency is 3.0×10^{14} Hz?
   ($h = 6.63 \times 10^{-34}$ J•s; 1 eV = 1.60×10^{-19} J)

ANS:
1.2 eV

Holt Physics Assessment Item Listing
245

2. What is the energy of a photon whose frequency is 5.0×10^{14} Hz? ($h = 6.63 \times 10^{-34}$ J•s; 1 eV = 1.60×10^{-19} J)

ANS:
2.1 eV

3. What is the energy of a photon whose frequency is 7.0×10^{14} Hz? ($h = 6.63 \times 10^{-34}$ J•s; 1 eV = 1.60×10^{-19} J)

ANS:
2.9 eV

4. What is the energy of a photon whose frequency is 9.6×10^{14} Hz? ($h = 6.63 \times 10^{-34}$ J•s; 1 eV = 1.60×10^{-19} J)

ANS:
4.0 eV

5. What is the energy of a photon whose frequency is 8.2×10^{14} Hz? ($h = 6.63 \times 10^{-34}$ J•s; 1 eV = 1.60×10^{-19} J)

ANS:
3.4 eV

6. What is the de Broglie wavelength for a proton that has a mass of 1.67×10^{-27} kg and is moving at a speed of 1.3×10^3 m/s? ($h = 6.63 \times 10^{-34}$ J•s)

ANS:
3.0×10^{-10} m

7. What is the de Broglie wavelength for a proton that has a mass of 1.67×10^{-27} kg and is moving at a speed of 2.7×10^5 m/s? ($h = 6.63 \times 10^{-34}$ J•s)

ANS:
1.5×10^{-12} m

8. What is the de Broglie wavelength for a proton that has a mass of 1.67×10^{-27} kg and is moving at a speed of 5.2×10^5 m/s? ($h = 6.63 \times 10^{-34}$ J•s)

ANS:
7.63×10^{-13}

9. What is the de Broglie wavelength for a proton that has a mass of 1.67×10^{-27} kg and is moving at a speed of 1.1×10^4 m/s? ($h = 6.63 \times 10^{-34}$ J•s)

ANS:
3.6×10^{-11} m

10. What is the de Broglie wavelength for a proton that has a mass of 1.67×10^{-27} kg and is moving at a speed of 9.9×10^6 m/s? ($h = 6.63 \times 10^{-34}$ J•s)

ANS:
4.0×10^{-14} m

MULTIPLE CHOICE

1. Which of the following exhibits zero resistance below a critical temperature?
    a. insulator
    b. conductor
    c. semiconductor
    d. superconductor

    ANS: C

2. Which of the following exhibits very high resistance?
    a. insulator
    b. conductor
    c. semiconductor
    d. superconductor

    ANS: A

3. Which of the following exhibits properties of both a conductor and a resistor?
    a. glass rod
    b. copper wire
    c. semiconductor
    d. superconductor

    ANS: C

4. Electrons are NEVER free to move within a solid and carry electric current in a(n)
    a. insulator.
    b. conductor.
    c. semiconductor.
    d. superconductor.

    ANS: A

5. Which statement is NOT correct?
    a. Valence electrons are electrons in the outermost shell of an atom.
    b. The behavior of valence electrons determines the chemical properties of an atom.
    c. Valence electrons are tightly bound to the nucleus of an atom.
    d. Valence electrons can strongly interact with other atoms.

    ANS: C

6. Which of the following is likely to be a valence electron?
    a. the innermost electron of a mercury atom
    b. an electron in a cobalt atom in the second-to-outermost energy level
    c. an electron in a copper atom in the outermost energy level
    d. an electron in a uranium atom in the third-to-outermost energy level

    ANS: C

7. A valence electron is an electron
    a. in the outermost shell of an atom.
    b. that is no longer bound to the atom.
    c. in an atom's excited state.
    d. in an atom's ground state.

    ANS: A

8. When identical atoms are far apart, they have _____ energy-level diagrams and _____ wave functions.
   a. split; split
   b. identical; identical
   c. different; different
   d. level; split

   ANS: B

9. Which solid has a valence band that overlaps the conduction band?
   a. conductor
   b. semiconductor
      c insulator
   c. W
   d. superconductor

   ANS: A

10. Which solid has an empty conduction band and a full valence band?
    a. conductor
    b. semiconductor
    c. insulator
    d. superconductor

    ANS: C

11. Which solid has a full valence band and a small energy gap?
    a. conductor
    b. semiconductor
    c. insulator
    d. superconductor

    ANS: B

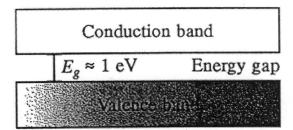

12. Which solid do the bands in the figure above represent?
    a. semiconductor
    b. nonconductor
    c. insulator
    d. superconductor

    ANS: A

Conduction band

$E_g \approx 5$ eV   Energy gap

Valence band

13. Which solid do the bands in the figure above represent?
    a. semiconductor   c. insulator
    b. conductor   d. superconductor

    ANS: C

14. An energy level that is not occupied by an electron in a semiconductor is a
    a. vacancy   c. diode.
    b. valance band.   d. hole.

    ANS: D

15. The motion of holes in a semiconductor
    a. is always in a direction opposite the motion of electrons.
    b. is always in the same direction as the motion of electrons.
    c. varies with the type of semiconductor.
    d. reverses at regular intervals.

    ANS: A

16. In a semiconductor containing only one element or compound,
    a. the number of conduction electrons is always greater than the number of holes.
    b. there is an equal number of conduction electrons and holes.
    c. the number of conduction electrons is always less than the number of holes.
    d. there is no relationship between the number of conduction electrons and holes.

    ANS: B

17. Charge can move in a substance
    a. only in the form of electrons.
    b. only as positively charged holes in an electron-poor substance.
    c. either in the form of electrons or as positively charged holes in an electron-poor
        substance.
    d. only in the valence band.

    ANS: C

18. When adding an impurity to a semiconductor in the process of doping, the impurity added
    a. must have extra valence electrons compared with the atoms in the intrinsic semiconductor.
    b. must have fewer valence electrons compared with the atoms in the intrinsic semiconductor.
    c. can have either extra valence electrons or fewer valence electrons compared with the atoms in the intrinsic semiconductor.
    d. must have the same number of valence electrons as the atoms in the intrinsic semiconductor

    ANS: C

19. If a silicon semiconductor, which has four valence electrons, is doped with atoms containing three valence electrons,
    a. holes can move throughout the material, even if no electrons are in the conduction band.
    b. the charge carriers are electrons having negative charges.
    c. positive holes that form in the donor level do not move easily.
    d. a donor atom occupies an energy level that lies just below the conduction band.

    ANS: A

20. When an n-type semiconductor and a p-type semiconductor are brought together to form a p-n junction, electrons from the _____ layer diffuse toward the _____ layer, and positive holes are formed in the _____ layer.
    a. n; p; n   c. p; n; n
    b. p; n; p   d. n; p; p

    ANS: A

21. Which statement about a p-n junction is NOT correct?
    a. It will allow current to flow in a circuit in either direction.
    b. The holes and electrons on either side of the p-n junction can migrate across the junction.
    c. Electrons from the n side nearest the junction diffuse toward the p side.
    d. On the p side nearest the junction, electrons diffuse so that holes move to the inside.

    ANS: A

22. At a p-n junction, when a free-moving electron and a free-moving hole meet,
    a. mobile charge carriers form.
    b. an internal electric field causes further diffusion of electrons and holes.
    c. the charges cancel each other, leaving no charge carriers.
    d. the depletion region continuously grows larger.

    ANS: C

23. If a large enough positive external voltage is applied to the p side of a diode,
    a. the electric potential on the negative side is raised compared with the positive side.
    b. the electric potential on the positive side is raised compared with the negative side.
    c. the diode is said to be reverse-biased.
    d. negative charge carriers cannot move.

    ANS: B

24. A p-n junction acts much like a
    a. diode.   c. transistor.
    b. generator.   d. transducer.

    ANS: A

25. If an ac current is sent through a diode, only the _____ component of the current is
    transmitted. This is why a diode is a _____.
    a. directional; capacitor   c. positive; rectifier
    b. nonzero; resistor   d. negative; rectifier

    ANS: C

26. The process that converts an alternating current into a direct current using a diode is
    a. insulating.   c. doping.
    b. superconducting.   d. rectification.

    ANS: D

27. When a diode is reverse-biased,
    a. the positive part of the original signal is suppressed.
    b. it has a very large resistance.
    c. the direct current produced is constant.
    d. an alternating current is produced.

    ANS: B

28. A diode rectifies an alternating current into a
    a. reverse-biased current.   c. constant direct current.
    b. forward-biased current.   d. pulsed direct current.

    ANS: D

29. Which of the following designates a junction transistor?
    a. n-p-p   c. n-n-p
    b. p-n-p   d. p-p-n

    ANS: B

30. Which statement about a p-n-p transistor is NOT correct?
    a. A very narrow n region is sandwiched between two p regions.
    b. It contains two p-n junctions.
    c. The base is more heavily doped than the emitter.
    d. The transistor has three leads that can amplify a signal.

    ANS: C

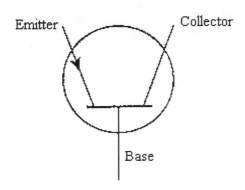

Emitter   Collector

Base

31. The circuit symbol shown is for a(n)
    a. n-p-n transistor.
    b. p-n-p transistor.
    c. rectifier.
    d. diode.

    ANS: B

32. Which statement about junction transistors is NOT correct?
    a. If the transistor is properly biased, it acts as a current amplifier.
    b. The output current is directly proportional to the input current.
    c. Two batteries are necessary to properly bias the transistor.
    d. The collector current is inversely proportional to the base current.

    ANS: D

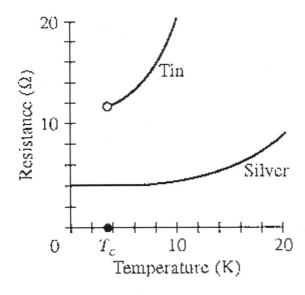

33. What can be determined from the graph above?
    a. Silver becomes a superconductor above temperature T_C
    b. Tin becomes a superconductor at temperature T_C.
    c. Tin never becomes a superconductor.
    d. The resistance of silver goes to zero at temperature T_C.

    ANS: B

Holt Physics Assessment Item Listing
253

34. Which statement is NOT supported by the graph above?
    a. At a temperature of about 10 K, the resistance of silver increases.
    b. Tin becomes a superconductor just below a temperature of T_C.
    c. The resistance of tin goes to zero at temperature T_C.
    d. The resistance of silver goes to zero at a temperature of 0 K.

    ANS: D

35. Some conductors exhibit nonzero resistance even at absolute zero because
    a. their atomic vibrations have too large an amplitude.
    b. the resistance steadily decreases as the substance is cooled.
    c. the geometric pattern of the crystal in a solid is flawed.
    d. the atoms of the lattice are lined up perfectly.

    ANS: C

36. Which statement about superconductors is correct?
    a. Superconductors exhibit nonzero resistance because of lattice imperfection.
    b. The resistance of a superconductor goes to zero below a certain nonzero temperature.
    c. Even in the absence of thermal vibrations, a superconductor exhibits a residual resistance.
    d. At its critical temperature, a superconductor's resistance suddenly increases

    ANS: B

37. If a substance is superconducting, it is likely to
    a. contain lattice imperfections.
    b. have a flawed crystal pattern.
    c. have a regular, geometric pattern of its crystal.
    d. have imperfect geometric arrangement of its atoms.

    ANS: C

38. According to the BCS theory, two electrons that travel through the lattice and behave like a single particle are a(n)
    a. electron pair
    b. Bardeen pair.
    c. Cooper pair
    d. Schrieffer pair.

    ANS: C

39. According to the BCS theory, if one electron scatters, the other electron in a pair
    a. closely follows the path of the scattered electron, keeping the total momentum constant.
    b. does not scatter, keeping the total momentum constant.
    c. is not affected.
    d. also scatters, keeping the total momentum constant.

    ANS: D

SHORT ANSWER

1. Why is it easy for charge carriers to move in a good conductor but not in an insulator?

   ANS:
   It is easy for charge carriers to move in a good conductor because the conductor has low resistance, and an insulator has a very high resistance.

2. Why are energy bands important in conducting electricity?

   ANS:
   The conduction of electricity depends on the arrangement of the electrons in the bands because a moving electron can move only to an unfilled final energy level.

3. As identical atoms are brought close together, their wave functions _____.

   ANS:
   overlap

4. As identical atoms are brought close together, their energy levels _____ and become energy bands.

   ANS:
   split

5. The highest band containing occupied energy levels is known as the

   ANS:
   valence band

6. Why is the energy level in an atom altered by the influence of the electric field of another atom?

   ANS:
   Because no two electrons in the same system can occupy the same state.

7. The minimum energy separation between the highest occupied state and the lowest energy state is called a(n) _____. An electron in an insulator or semiconductor cannot have a value for its energy that would lie within this energy separation.

   ANS:
   band gap (or energy gap)

8. Whether the _____ is completely full or only partially full is important in determining the electrical properties of a material.

   ANS:
   valence band (or highest energy band)

9. A semiconductor that moves an electric current by allowing electrons to flow freely is a(n)_____-type semiconductor. A semiconductor that moves electric current by allowing positive charges (holes) to move throughout the material is a(n) _____-type semiconductor.

ANS:
n; p

10. An n-type semiconductor consists of silicon or germanium and an impurity atom, called a(n) _____. This impurity alters the band structure of the solid so that the extra electron occupies an energy, or _____, that lies just below the conduction band.

ANS:
donor atom; donor level

11. A p-type semiconductor consists of a main semiconductor material that has four valence electrons and a trace of an element called a(n) _____, which has three valence electrons. When admitted to the crystal structure, this impurity creates a(n) _____.

ANS:
acceptor; hole

12. A p-n junction diode is connected to a voltage source so that the positive end of the voltage source is connected to the p end of the diode and the negative end of the source is connected to the n end of the diode. Does the circuit conduct the current in this case?

ANS:
yes

13. At the emitter-base junction of a p-n-p transistor, how do holes move from the p-type emitter to the base?

ANS:
The heavily doped emitter is forward biased so that positive charge carriers (the holes) readily flow through the emitter terminal to the base.

14. Do holes recombine with electrons in the base? Explain your answer.

ANS:
Most of the holes do not recombine with electrons in the base because the base is very narrow.

15. How can current be reduced as it flows from the emitter through the base to the collector?

ANS:
Positive charge carriers (holes) that recombine in the base accumulate and prevent holes from flowing in.

16. How do holes accelerate across the reverse-biased base-collector junction of a transistor?

ANS:
The barrier has an opposite effect on the holes because they are on the n side of the junction, where the charge carriers are usually electrons.

17. What is the important feature of the BCS theory of superconductivity?

ANS:
It provides a new understanding of the way that electrons, traveling in pairs, move through the lattice of a superconductor.

18. What idea does the BCS theory of superconductivity describe?

ANS:
Electrons collide in a superconductor, but the electrons do not alter the total momentum of the pair of electrons. It seems that the electrons move unimpeded through the lattice.

19. Using superconducting wires in an electromagnet solves what problem?

ANS:
Electromagnets dissipate a great amount of energy because of their electrical resistance to current flow. Superconducting wires eliminate this energy loss.

20. What is the advantage of using superconducting magnets in magnetic resonance imaging (MRI)?

ANS:
MRI uses relatively safe radio-frequency radiation rather than high-energy X rays to produce images of body sections.

21. What is the advantage of using superconducting film to connect computer chips?

ANS:
It increases the speed of computer operation.

22. What is the advantage of using a superconducting ring to store electric current or energy?

ANS:
There is no resistance in a superconducting ring; therefore, a current introduced in this type of ring will continue indefinitely and could be extracted at any later time.

MULTIPLE CHOICE

1. The nucleus of an atom is made up of
    a.   electrons and protons.
    b.   electrons and neutrons..
    c.   protons, electrons, and neutrons
    d.   protons and neutrons

    ANS: D   DIF: I   OBJ: 25-1.1

2. The atomic number of a given element is equivalent to
    a.   the proton number in the nucleus.
    b.   the neutron number in the nucleus.
    c.   the sum of the protons and neutrons in the nucleus.
    d.   the number of electrons in the outer shells.

    ANS: A   DIF: I   OBJ: 25-1.1

3. Rutherford's experiments involving the use of alpha particle beams directed onto thin metal foils demonstrated the existence of which of the following?
    a.   neutron
    b.   proton
    c.   nucleus
    d.   positron

    ANS: C   DIF: l: OBJ: 25-1.1

4. The mass number of a nucleus is the
    a.   number of neutrons present.
    b.   number of protons present.
    c.   difference in neutron and proton numbers.
    d.   sum of neutron and proton numbers.

    ANS: D   DIF: I   OBJ: 25-1.1

5. If there are 128 neutrons in Pb-210, how many neutrons are found in the nucleus of Pb-206?
    a.   122
    b.   124
    c.   126
    d.   130

    ANS: B   DIF: I   OBJ: 25-1.1

6. The binding energy of a nucleus is
    a.   the energy needed to remove one of the nucleons.
    b.   the average energy with which any nucleon is bound in the nucleus.
    c.   the energy released when nucleons bind together to form a stable nucleus.
    d.   the mass of the nucleus times c^2.

    ANS: C   DIF: II   OBJ: 25-1 .2

7. If the stable nuclei are plotted with neutron number versus proton number, the curve formed by the stable nuclei does not follow the line $N = Z$. This is predicted by examining how the binding energy is influenced by
    a.   the volume of the nucleus.
    b.   the size of the nuclear surface.
    c.   the Coulomb repulsive force.
    d.   the proton-neutron mass difference.

    ANS: C   DIF: II   OBJ: 25-1.2

8. The more protons in the nucleus, the _____ the repulsive force. Therefore, _____ neutrons are needed to keep the nucleus stable.
   a. stronger; more
   b. stronger; less
   c. weaker; more
   d. stronger; no

   ANS: A   DIF: I   OBJ: 25-1.2

9. Light nuclei are stable when
   a. they contain more protons than neutrons.
   b. they contain more neutrons than protons.
   c. they contain equal numbers of protons and neutrons.
   d. the Coulomb force is stronger than the nuclear force.

   ANS: C   DIF: I   OBJ: 25-1.2

10. Heavy nuclei are stable when
    a. they contain more protons than neutrons.
       b). hey contain more neutrons than protons.
    b. H
    c. they contain equal numbers of protons and neutrons.
    d. the Coulomb force is stronger than the nuclear force.

    ANS: B   DIF: I   OBJ: 25-1.2

11. What is the binding energy per nucleon of $^{197}_{79}$Au? ($c^2 = 931.50$ MeV/u; atomic masses: $^{197}_{79}$Au $= 196.966\ 543$ u; $^{1}_{1}$H $= 1.007\ 825$ u; $m_n = 1.008\ 665$ u)
    a. 7.4551 MeV/nucleon
    b. 12.4512 MeV/nucleon
    c. 7.9159 MeV/nucleon
    d. 12.8764 MeV/nucleon

    ANS: C   DIF: IIIB   OBJ: 25-1.3

12. What is the binding energy per nucleon of the tritium nucleus, $^{3}_{1}$H? ($c^2 = 931.50$ MeV/u; atomic masses: $^{3}_{1}$H $= 3.016\ 049$ u; $^{1}_{1}$H $= 1.007\ 825$ u; $m_n = 1.008\ 665$ u)
    a. 2.243 MeV/nucleon
    b. 2.454 MeV/nucleon
    c. 2.827 MeV/nucleon
    d. 2.196 MeV/nucleon

    ANS: C   DIF: IIIB   OBJ: 25-1.3

13. Calculate the binding energy of the carbon-12 nucleus. ($c^2 = 931.50$ MeV/u; atomic masses: $^{12}_{6}$C $= 12.000\ 000$ u; $^{1}_{1}$H $= 1.007\ 825$ u; $m_n = 1.008\ 665$ u)
    a. 26.880 MeV
    b. 44.554 MeV
    c. 81.972 MeV
    d. 92.163 MeV

    ANS: D   DIF: IIIC   OBJ: 25-1.3

14. Calculate the binding energy of the $^{16}_{8}$O nucleus. ($c^2 = 931.50$ MeV/u; atomic masses: $^{16}_{8}$O = 15.994 915 u; $^{1}_{1}$H = 1.007 825 u; $m_n = 1.008$ 665 u)
   a.   127.62 MeV
   b.   63.562 MeV
   c.   51.981 MeV
   d.   39.410 MeV

   ANS: A   DIF: IIIC   OBJ: 25-1.3

15. How does a radioactive isotope that emits an alpha particle change?
   a.   atomic number decreases by four one
   b.   mass number decreases by four
   c.   atomic number decreases by
   d.   mass number decreases by one

   ANS: B   DIF: I   OBJ: 25-2.1

16. Of the main types of radiation emitted from naturally radioactive isotopes, which is the most penetrating?
   a.   alpha
   b.   beta
   c.   gamma
   d.   positron

   ANS: C   DIF: I   OBJ: 25-2.1

17. The components of natural radiation, in order from least to most penetrating, are
   a.   alpha, beta, and gamma.
   b.   gamma, beta, and alpha.
   c.   beta, gamma, and alpha.
   d.   alpha, gamma, and beta.

   ANS: A   DIF: I   OBJ: 25-2.1

18. The alpha emission process results in the daughter nucleus differing in what manner from the parent?
   a.   atomic mass increases by one
   b.   atomic number decreases by two
   c.   atomic number increases by one
   d.   atomic mass decreases by two

   ANS: B   DIF: I   OBJ: 25-2.1

19. The beta emission process results in the daughter nucleus differing in what manner from the parent?
   a.   atomic mass changes by one
   b.   atomic number changes by two
   c.   atomic number changes by one
   d.   atomic mass changes by two

   ANS: C   DIF: I   OBJ: 25-2.1

20. In this radioactive formula, what does X represent? $^{220}_{86}$Rn → $^{216}_{84}$Po + X
   a. $^{0}_{-1}e$
   b. $^{0}_{1}e$
   c. λ
   d. $^{4}_{2}$He

   ANS: D   DIF: II   OBJ: 25-2.2

21. What particle is emitted when Pu-240 decays to U-236? (atomic numbers of Pu = 94 and U = 92)
    a. alpha
    b. beta
    c. positron
    d. gamma

    ANS: A   DIF: I   OBJ: 25-2.2

22. What particle is emitted when P-32 decays to S-32? (atomic numbers of P = 15 and S = 16)
    a. alpha
    b. beta
    c. positron
    d. gamma

    ANS: B   DIF: I   OBJ: 25-2.2

23. When bismuth-214 emits a beta particle, the remaining daughter nucleus is
    a. lead-213.
    b. actinium-215.
    c. polonium-214
    d. bismuth-215.

    ANS: C   DIF: II   OBJ: 25-2.2

24. Radium-226 decays to radon-222 by emitting
    a. beta particles.
    b. alpha particles.
    c. gamma particles.
    d. positrons.

    ANS: B   DIF: I   OBJ: 25-2.2

25. A radioactive material initially is observed to have an activity of 800 counts/s. If 4 h later it is observed to have an activity of 200 counts/s, what is its half-life?
    a. 1 h
    b. 2 h
    c. 4 h
    d. 8 h

    ANS: B   DIF: IIIA   OBJ: 25-2.3

26. If a fossil bone is found to contain one-eighth as much carbon-14 as the bone of a living animal, what is the approximate age of the fossil? (half-life of carbon-14 = 5800 years)
    a. 8000 years
    b. 17 400 years
    c. 23 000 years
    d. 46 400 years

    ANS: B   DIF: IIIB   OBJ: 25-2.3

27. Samples of two different isotopes, X and Y, both contain the same number of radioactive atoms. Sample X has a half-life twice that of Y. How do their rates of radiation compare?
    a. X has a greater rate than Y.
    b. X has a smaller rate than Y.
    c. Rates of X and Y are equal.
    d. Rate depends on atomic number, not half-life.

    ANS: B   DIF: II   OBJ: 25-2.3

28. A pure sample of Ra-226 contains 2.0×10^{14} atoms of the isotope. If the half-life of Ra-226 is 1.6×10^3 years, what is the decay rate of this sample?
    a. 6.7×10^9 decays/year
    b. 8.7×10^{-10} decays/year
    c. 9.4×10^{-10} decays/year
    d. 13×10^{-10} decays/year

    ANS: B   DIF: IIIB   OBJ: 25-2.3

29. A pure sample of Ra-226 contains 2.0×10^{14} atoms of the isotope. If the half-life of Ra-226 = 1.6 $\times 10^3$ years, what is the decay rate of this sample?
    (1 Ci = 3.7×10^{10} decays/s)
    a. 2.7×10^{-12} Ci
    b. 3.4×10^{-10} Ci
    c. 7.4×10^{-8} Ci
    d. 9.6×10^{-6} Ci

    ANS: C   DIF: IIIC   OBJ: 25-2.3

30. Tritium has a half-life of 12.3 years. How many years will have elapsed when the radioactivity of a tritium sample has decreased to 10 percent of its original value?
    a. 31 years
    b. 41 years
    c. 84 years
    d. 123 years

    ANS: B   DIF: IIIB   OBJ: 25-2.3

31. In fission reactions, the binding energy per nucleon must _____ atomic number.
    a. remain constant; increasing
    b. remain constant; decreasing
    c. increase; increasing
    d. decrease; increasing

    ANS: D   DIF: I   OBJ: 25-3.1

32. In fusion reactions, the binding energy per nucleon must _____ with _____ atomic number.
    a. remain constant; increasing
    b. remain constant; decreasing
    c. increase; increasing
    d. decrease; increasing

    ANS: C   DIF: I   OBJ: 25-3.1

33. In order to adequately control a chain reaction, it is necessary to have within the fissionable material a nonfissionable material that _____ neutrons.
    a. contributes
    b. absorbs
    c. emits
    d. reflects

    ANS: B   DIF: I   OBJ: 25-3.2

34. An uncontrolled chain reaction begins when two or more masses of fissionable material, each below the _____, are brought together very quickly and when there are no _____ -absorbing materials present.
    a. melting point; light
    b. critical mass; neutron
    c. ionization energy; proton
    d. ground state; electron

    ANS: B   DIF: I   OBJ: 25-3.2

35. How is a fission reactor different from a fusion reactor?
    a. The fuel is cheaper.
    b. The fuel must be processed.
    c. There is less radioactive waste.
    d. The transportation of fuel is safer.

    ANS: B   DIF: I   OBJ: 25-3.3

36. A thermonuclear fusion reaction cannot be maintained in the oceans of Earth because
    a. the temperature is not high enough.
    b. the density is not high enough.
    c. there is insufficient deuterium in the ocean.
    d. the deuterium in the ocean is not radioactive.

    ANS: A   DIF: I   OBJ: 25-3.3

37. Which interaction of nature is weakest?
    a. strong
    b. weak
    c. electromagnetic
    d. gravitational

    ANS: D   DIF: I   OBJ: 25-4.1

38. Which interaction of nature binds neutrons and protons into nuclei and is the strongest?
    a. strong
    b. weak
    c. electromagnetic
    d. gravitational

    ANS: A   DIF: I   OBJ: 25-4.1

39. Which interaction of nature depends on the distance through which it acts for its strength and is involved in beta decay?
    a. strong
    b. weak
    c. electromagnetic
    d. gravitational

    ANS: B   DIF: I   OBJ: 25-4.1

40. Which interaction of nature holds the planets, stars, and galaxies together, even though its effect on elementary particles is negligible?
    a. strong
    b. weak
    c. electromagnetic
    d. gravitational

    ANS: D   DIF: I   OBJ: 25-4.1

41. Which interaction of nature binds atoms and molecules by attracting unlike charges and repulsing like charges?
    a. strong
    b. weak
    c. electromagnetic
    d. gravitational

    ANS: C   DIF: I   OBJ: 25-4.1

42. Which of the following do physicists believe are fundamental particles?
    a. three quarks and three leptons
    b. six quarks and three leptons
    c. three quarks and six leptons
    d. six quarks and six leptons

    ANS: D   DIF: I   OBJ: 25-4.2

43. What is the charge on all quarks?
    a. $\frac{1}{3}e$ or $\frac{2}{3}e$
    b. $\frac{1}{3}e$ or $-\frac{2}{3}e$
    c. $-\frac{1}{3}e$ or $\frac{2}{3}e$
    d. $-\frac{1}{3}e$ or $-\frac{2}{3}e$

    ANS: C   DIF: I   OBJ: 25-4.2

44. Which statement about quarks is NOT correct?
    a. Only two quarks are needed to construct a hadron.
    b. An isolated quark has been observed by physicists.
    c. Every quark has an antiquark of opposite charge.
    d. There are six quarks that fit together in pairs.

    ANS: B   DIF: I   OBJ: 25-4.2

45. Which of the following is an example of a baryon?
    a. meson   c. lepton
    b. electron   d. proton and neutron

    ANS: A   DIF: I   OBJ: 25-4.2

46. Hadrons are composed of
    a. leptons.   c. mesons, baryons, and antibaryons.
    b. electrons.   d. neutrinos.

    ANS: C   DIF: I   OBJ: 25-4.2

47. During the radiation era of the big bang theory, most energy was in the form of radiation. What form did matter take?
    a. hydrogen atoms   c. electrons
    b. ions   d. ions and electrons

    ANS: D   DIF: I   OBJ: 25-4.3

SHORT ANSWER

1. _____ are atoms that have the same atomic number but different neutron numbers.

   ANS:
   Isotopes

   DIF: I   OBJ: 25-1.1

2. Why do elements containing more than 83 protons have unstable nuclei?

   ANS:
   The repulsive forces between protons cannot be compensated by the addition of more neutrons.

   DIF: I   OBJ: 25-1 .2

3. List alpha, beta, and gamma radiations in order of decreasing speed.

   ANS:
   gamma particles, beta particles, alpha particles

   DIF: I   OBJ: 25-2.1

4. Use alternative terms to identify the composition of alpha particles, beta particles, and gamma particles.

ANS:
two protons, two neutrons; electrons or positrons; photons

DIF: I   OBJ: 25-2.1

5. What does the value of λ indicate for any isotope?

ANS:
the rate at which that isotope decays

DIF: I   OBJ: 25-2.3

6. What is a half-life?

ANS:
the time required for half the original nuclei of a radioactive material to undergo radioactive decay

DIF: I   OBJ: 25-2.3

7. Complete the following nuclear reaction, and state whether it is a fusion or fission reaction.
$$^{230}_{90}\text{Th} \rightarrow {}^{226}_{88}\text{Ra} + \underline{\hspace{2cm}}$$

ANS:
$^{4}_{2}\text{He}$; fission reaction

DIF: IIIA   OBJ: 25-3.1

8. Complete the following nuclear reaction, and state whether it is a fusion or fission reaction.
$$^{22}_{10}\text{Ne} + {}^{4}_{2}\text{He} \rightarrow \underline{\hspace{2cm}} + 2\,{}^{1}_{0}\text{n}$$

ANS:
$^{24}_{12}\text{Mg}$; fusion reaction

DIF: IIIA   OBJ: 25-3.1

9. What is nuclear fission?

ANS:
Nuclear fission is a process during which a heavy nucleus splits into two or more lighter nuclei. When it occurs naturally, the nucleus releases energy.

DIF: I   OBJ: 25-3.1

10. What is nuclear fusion?

ANS:
Nuclear fusion is a process during which two or more nuclei combine to form a heavier nucleus.

DIF: I   OBJ: 25-3.1

11. In the nuclear chain reaction of uranium-235, what particle reacts with the uranium nucleus to become a product of the reaction?

ANS:
neutron

DIF: II   OBJ: 25-3.2

12. What can trigger a chain reaction in a nuclear reactor?

ANS:
Released neutrons can be captured by other nuclei, making these nuclei unstable. This triggers additional fission events and, possibly, a chain reaction.

DIF: I   OBJ: 25-3.2

13. Why are fusion reactors a desirable source of energy?

ANS:
They use deuterium as a fuel source, and deuterium is commonly found in sea water, which is cheap and plentiful. Also, they produce less waste compared with fission reactors.

DIF: I   OBJ: 25-3.3

14. In a fission reactor, what must occur when uranium-238 absorbs neutrons instead of undergoing fission?

ANS:
Reactor fuels must be processed or enriched to increase the proportion of ^{235}U to a level that the reaction is able to sustain.

DIF: II   OBJ: 25-3.3

15. How is particle physics related to the model of the universe?

ANS:
Particle physics helps us understand the origin and evolution of the universe. According to the big bang theory, particles (and matter and energy) evolved similar to the four interactions of physics.

DIF: I   OBJ: 25-4.3

16. According to the big bang theory, what occurred in the brief instant after the big bang?

ANS:
The four fundamental interactions of physics operated in a unified manner. The high temperatures and energy caused all particles and energy to be indistinguishable.

DIF: I   OBJ: 25-4.3

PROBLEM

1. Calculate the binding energy of the iron-56 nucleus. ($c^2 = 931.50$ MeV/u; atomic masses:

    $_{26}^{56}$Fe = 55.934 940 u; $_{1}^{1}$H = 1.007 825 U; m_n = 1.008 665 u)

    ANS:
    492.26 MeV

    DIF: IIIC   OBJ: 25-1.3

2. Calculate the binding energy of the potassium-39 nucleus. ($c^2 = 931.50$ MeV/u; atomic masses: =

    38.963 708 u; $_{1}^{1}$H = 1.007 825 U; m_n = 1.008 665 u)

    ANS:
    333.72 MeV

    DIF: IIIC   OBJ: 25-1.3

3. Calculate the binding energy of the chlorine-35 nucleus. ($c^2 = 931.50$ MeV/u; atomic masses:

    $_{17}^{35}$Cl = 34.968 853 u; $_{1}^{1}$H = 1.007 825 u; , m_n = 1.008 665 u)

    ANS:
    298.21 MeV

    DIF: IIIC   OBJ: 25-1.3

4. Calculate the binding energy of the sulfur-32 nucleus. ($c^2 = 931.50$ MeV/u; atomic masses:

    $_{16}^{32}$S = 31.972 071 u; $_{1}^{1}$H = 1.007 825 u; m_n = 1.008 665 u)

    ANS:
    271.78 MeV

    OBJ: 25-1.3

5. Calculate the binding energy of the phosphorus-31 nucleus. ($c^2 = 931.50$ MeV/u; atomic masses: $^{31}_{15}P = 30.973\ 762$ U; $^{1}_{1}H = 1.007\ 825$ u; $m_n = 1.008\ 665$ u)

ANS:
262.92 MeV

DIF: IIIC   OBJ: 25-1.3

6. Calculate the binding energy of the aluminum-27 nucleus. ($c^2 = 931.50$ MeV/u; atomic masses: $^{27}_{13}Al = 26.981\ 534$ u; $^{1}_{1}H = 1.007\ 825$ U; $m_n = 1.008\ 665$ u)

ANS:
146.27 MeV

DIF: IIIC   OBJ: 25-1.3

7. Calculate the binding energy of the sodium-23 nucleus. ($c^2 = 931.50$ MeV/u; atomic masses: $^{23}_{11}Na = 22.989\ 767$ u; $^{1}_{1}H = 1.007\ 825$ U; $m_n = 1.008\ 665$ u)

ANS:
186.56 MeV

DIF: IIIC   OBJ: 25-1.3

8. Calculate the binding energy of the copper-63 nucleus. ($c^2 = 931.50$ MeV/u; atomic masses: $^{63}_{29}Cu = 62.929\ 599$ u; $^{1}_{1}H = 1.007\ 825$ u; $m_n = 1.008\ 665$ u)

ANS:
551.38 MeV

DIF: IIIC   OBJ: 25-1.3

9. Calculate the binding energy of the zinc-64 nucleus. ($c^2 = 931.50$ MeV/u; atomic masses: $^{64}_{30}Zn = 63.929\ 144$ u; $^{1}_{1}H = 1.007\ 825$ u; $m_n = 1.008\ 665$ u)

ANS:
559.10 MeV

DIF: IIIC   OBJ: 25-1.3

10. Calculate the binding energy of the cobalt-59 nucleus. ($c^2 = 931.50$ MeV/u; atomic masses:

$^{59}_{27}\text{Co} = 58.933\ 198$ u; $^{1}_{1}\text{H} = 1.007\ 825$ u; $m_n = 1.008\ 665$ u)

ANS:
517.30 MeV

DIF: IIIC   OBJ: 25-1.3

GREYSCALE

BIN TRAVELER FORM

Cut By _Dhiyk Aldina_ Qty 53 Date 3-11-24

Scanned By _____ Qty _____ Date _____

Scanned Batch IDs

_____ _____ _____

Notes / Exception